BEES
AND
BEEKEEPING
AN EDUCATIONAL BOOK
FOR
HIGH SCHOOL AGE GROUPS

by
Jim Wright
2019

BEES AND BEEKEEPING © Jim Wright

All rights reserved. No part of this publication may be reproduced, stored in a retrieval system, transmitted in any form or by any means electronic, mechanical, including photocopying, recording or otherwise without prior consent of the copyright holders.

ISBN 978-1-912271-57-3

Published by Northern Bee Books, 2020
Scout Bottom Farm
Mytholmroyd
Hebden Bridge HX7 5JS (UK)

Design and artwork by DM Design and Print

BEES
AND
BEEKEEPING
AN EDUCATIONAL BOOK
FOR
HIGH SCHOOL AGE GROUPS

PREFACE

Is there a need for a book about bees and beekeeping for high school ages? Experience talking to school children of all ages at shows, particularly the Tocal Agricultural College Field Days, suggests that there is. The response of teen agers to a display of live bees and exposure to information about bees and beekeeping varies from keen interest to a shrug with provocative comments such as, "Arrgh, honey is just bee vomit", or an aggressive question, "What would you do if I let the bees out ?" Such typical teenage aggressive behaviour may well be just a front hiding a fertile soil for education about these fascinating creatures and their importance to us.

A separate work is directed to young children, and much of this is repeated here, somewhat modified, as a basic introduction.

The focus here turns to the evolution of bees, their place in the animal kingdom and their anatomy and function. With the assumption that these older age groups have a basic understanding of the biology of sexual reproduction, this subject is addressed in some detail.

Emphasis is placed on the irreplaceable role played by insects and bees in our environment and our food sources. Problems of beekeeping, amateur and commercial, are addressed with focus on diseases and pests. The age group targeted by this work has a vital interest in the world-wide threat to bees and this subject is afforded some prominence.

In the hope of inspiring continued interest in this important subject, readers are invited to consider contributing to the health of the environment by promoting bee friendly garden practices and maybe even becoming amateur beekeepers.

The early chapters provide a general introduction to bees, followed by their evolution and place in the animal kingdom, biology and diseases. Chapters are devoted to beekeepers and beekeeping, bee stings, honey, pollination and Australian native bees. Later chapters address matters of general interest.

So why a book when the internet now is the preferred information source, particularly of this age group? It is hoped and assumed that there will always be a need for a book to keep on the shelf and to curl up with in chair or bed.

TABLE OF CONTENTS

	PAGE
GLOSSARY	2
ABOUT INSECTS AND BEES	4
THE HONEY BEE	7
THE BEEHIVE	15
BEESWAX AND HONEYCOMB	18
EVOLUTION OF BEES	20
BIOLOGY OF BEES	21
Place of bees in the animal kingdom, Taxonomy	21
Anatomy	23
Reproductive biology of the honey bee	27
Swarming	33
Sociology; who controls the colony and how?	36
Physiological functions of bees	38
Nutrition, sight, hearing, smell, taste, walking, flying, navigation, communication	
DISEASES, PARASITES AND PESTS	45
The Asian Bee	50
BEE STINGS	53

BEEKEEPERS and BEEKEEPING 59
Harvesting Honey

Amateur and Professional Beekeeping

Management, caring for bees

Queen Breeding

MORE ABOUT HONEY 78

POLLINATION 83

BEEKEEPERS OF THE CLOTH: Monks, Priests and Clergy 86
The Patron Saint of Beekeeping 88

HOW WAS THE EUROPEAN BEE BROUGHT
TO AUSTRALIA? 89

THREATS TO BEES Are Bees disappearing? 90

MEMORIES OF A DYING BEE 93
A journey through the sequential duties in the lifetime of a bee 93

AUSTRALIAN NATIVE BEES 97

WOULD YOU LIKE TO BE A BEEKEEPER? 100

TRIVIA: Some statistical figures for fun 106

ACKNOWLEDGEMENTS 109

GLOSSARY

Antibiotic: A substance produced by a living organism which kills bacteria.

Antiseptic: A substance used in medicine applied externally to sterilise and kill bacteria.

Bacterium (plural "bacteria"): A microscopic organism, a "germ" of particular variety and shape.

Carbohydrate: Sugars and complex substances made up of carbon, hydrogen and oxygen molecules.

Colony: A unit of social bees consisting of one queen, drones and many worker bees.

DPI: New South Wales Department of Primary Industries, the government body responsible for beekeeping.

Fat: High energy organic material made up of "fatty acids", some essential for life. Used in animals to store energy and insulate against cold.

Hormone: A chemical substance formed in the body and influencing or controlling body functions. "A substance formed in an organ and serving to excite some vital process" (*Oxford Dictionary*). It works within the body in which it is produced. (*Contrast to "pheromone" below.*)

Hygroscopic: Drawing water from surroundings; honey is hygroscopic.

Neuron: A nerve cell with its attached nerve filaments.

Palynology: The study of pollen. Melissopalynology is the study of pollen grains in honey.

Parasite: "An animal or plant which lives in or upon another organism (its host) and draws its nutriment directly from it" (*Oxford English Dictionary*).

Pheromone: Similar to a hormone but produced by a creature, (eg a bee) to act on and influence the behavior of others in the society or colony. In bees this action is commonly by smell. Pheromones are produced by the queen to influence and control worker bees and pheromones are produced by worker bees to act upon other workers.

Prebiotic: A food substance which encourages the growth of "good" bacteria in the gut; thus helping to suppress pathogenic ("bad") germs. Honey is such a food.

Probiotic: A live culture of harmless bacteria eaten to thrive in the gut and suppress the growth of pathogens. Example: Lactobacillus acidophilus.

Protein: Body building material, contain nitrogen as well as carbon, hydrogen and oxygen. Made of amino acids some of which are essential to maintaining life.

Spore: A capsule-like body formed by bacteria as a dormant but potentially active form to perpetuate the species by resistance to destruction enabling it to remain viable for a long period.

Swarm: As a noun used synonymously with colony. As a verb "to swarm" it refers to the bee colony separating into two with one half of the bees swirling out of the hive to form a new colony; the process of "swarming".

Syndrome: A set of symptoms occurring together to cause or constitute a recognisable condition; not a specific disease or entity of itself.

Symbiotic: Living together, mutually supportive or dependent.

Virus: A sub-microscopic organism which lives within a host cell exploiting and influencing its metabolism. A virus cannot survive outside a host cell.

Vitamins: Chemical substances in animal biology essential in minute amounts to a particular metabolic process.

ABOUT INSECTS AND BEES

Bees are insects. What is an insect?

Insects are small creatures which populate the earth is huge numbers. There are more insects than all the other types of animal life put together, over half a million species. You know them, flies, mosquitoes, ants, cockroaches, and bees. Some insects, such as mosquitoes, are harmful to humans because they can carry disease. Others, like bees, are very important for our survival. Insects are the food of some birds and animals which is fortunate because they multiply so rapidly that they would soon cover the earth if enough were not eaten to keep the numbers down.

Flowering plants depend upon insects, especially bees, to fertilise them; and bees need flowers for their food. Flowers produce two important substances, sugary water (called nectar), and pollen which is a rich food for bees supplying their fat and protein. Insects, especially bees, visit flowers to collect this nectar and pollen for their own food and in doing so they spread pollen from flower to flower. This is called "pollination" and is necessary for plants to produce fruit and seeds. Bees and flowers depend on each other, they evolved together. We, as human beings, also depend on insects because most of our food is from crops grown as a result of this pollination.

Insects have three body parts, a head, a thorax or chest and abdomen. On the thorax are six legs, and (in some insects) wings.

They develop from eggs which hatch out, not into an adult, but into a grub, called a larva (pleural "larvae").

Recently laid eggs in the honeycomb.
Photograph supplied by DPI

You have seen many insect larvae in your garden and in trees; grubs of all shapes, sizes and colours, often well camouflaged like this green caterpillar on a green leaf.

Grubs eat large amounts of food and grow rapidly till they spin a coat for themselves called a cocoon. Now called a pupa, they sleep and while they are asleep they change slowly into the adult form, just like the adult that originally laid the egg, a bee, a moth, a butterfly, an ant, a fly. This process is called metamorphosis.

Stages in the development of insects

Egg   Larva   Pupa   Emerging adult

From pictures supplied by DPI

What sort of insect is a bee?

A bee is an insect which collects pollen from flowers for its own food and to feed it's young. Having developed, or evolved, with flowering plants, bees have been in the world for 200 million years, a thousand times longer than humans have been. There are over 20,000 species or varieties of bees in the world, and Australia has over 1,500 these. We have seen that they are essential for the growth of plants. Some make honey as their food, and one species, called the honey bee, makes so much of it that we can take some from them, called "robbing" or "harvesting", so we can enjoy its sweet flavour and the health benefits of this wonderful food. Before sugar cane was discovered in tropical countries and sugar was brought to Europe, honey was the universal sweetener for food.

THE HONEY BEE

The bee you most commonly see on flowers in your garden is called the "honey bee" because it makes the honey we enjoy. It is not the only bee that makes honey but is the only one that makes enough for us to take a substantial amount for ourselves. It is not native to Australia. It was brought here by ship from England 34 years after settlement at Sydney Cove, and the bees loved our gum trees.

Portrait of honey bee by Gina Cranson

Honey Bees busy on flowers.

Photograph by Judith Parker

Honey Bees and their Colonies

That bee you see in the picture above cannot live alone. It is called a social bee belonging to a large group called a colony or swarm containing between 10,000 and 80,000 bees. In that colony each bee has its job to do and each is dependent on others to do their work. A bee colony is like our community. Or like our bodies which are made up of thousands of small parts or cells which do not function alone but have an essential role in the physiology of the whole body.

In that colony there are three types or castes of bees each with separate functions.

First, the female, the **QUEEN:** There is only one queen to a colony, and she is the mother of all the other bees in it, thousands of them.

Photograph by Anthony Pyne

She is a little larger than the others. She has only two functions. The first is to lay eggs, more than a thousand eggs a day. Since bees live only about six weeks, the queen has to lay 1,000 or more eggs a day to produce enough new bees just to maintain the numbers. Her second duty is to smell, to exude an odour called a "pheromone" which spreads throughout the colony, controls the other bees and keeps them working as a huge team. She cannot feed herself or even clean herself. Other bees, like ladies in waiting, feed her, groom her and look after her.

They do not sting us and they can be handled without fear.

Queens have a sting, but it is not barbed and is used only to kill other queens. They do not sting us and they can be handled without fear.

Queens have a voice. They can emit a high pitched piping sound but it is rarely heard.

Today's fact

Queen bees emit a much larger range of sounds than normal bees

Queen bee surrounded by worker bees caring for her
Photograph by Des Cannon

This queen has been identified by the beekeeper with a number on her back. She is seen here with her tail end in a cell laying an egg

Photograph provided by Elizabeth Frost

Next, the **DRONES:** These are male bees, the boys, bigger than the others; big eyes, big shoulders, big hairy backside.

Photograph supplied by DPI

They have only one duty, and that is to mate with a young queen, and after they do that they die. Like the queen, they cannot feed themselves and must be cared for by the worker bees. In winter, when the colony is not breeding, the drones, no longer needed, are not fed and eventually they are cast out of the hive to die. Not a great life! Drones do not sting.

Now the **WORKER BEES:** These are the ones you see in your garden. They do all the work in the hive and in the field. They are all girls, but they are infertile females; they do not reproduce; only the queen does that. These are also the bees that sting. (Bees do not "bite", their jaws are very tiny, they sting with a barb in their tails.)

These worker bees live about six weeks and in that time they perform all the tasks required to keep the colony fed and healthy. They move from one task to another as they get older. First there's cleaning, the work of housemaids. Then they move to feeding babies (nursery maids), looking after the queen (ladies in waiting), packing honey and pollen in the honeycomb cells, (like packing the shelves in a supermarket). In the middle of their lives they make beeswax and honeycomb (builders). Later they act as air conditioning engineers, sitting at the hive entrance to blow air in and out to keep the air inside fresh and to cool it during hot weather. Then on to guard duty, (policewomen), they protect the colony from intruders. And their last duty is field work, called foraging, going out of the hive into trees and the flowers to collect nectar, pollen and water. That's why you see them on flowers and on your bird bath or at the side of your swimming pool. And all the while they are making honey from the nectar.

It is amazing that bees do all that, and never go to school or learning classes.

So when you see one of these bees lying on the footpath dying, don't be too sad for her; she is at the end of her life and has just worked herself to death. Bees work and fly till their wings become so tattered and torn that they can't manage to fly home. But beware! Don't touch a dying or drowning bee, she will sting!

Worker bees clustered on the front of a hive. Look closely and you will see little yellow dots on some of the bees. This is pollen collected from flowers.

Now let's look a little more closely at this foraging bee working on a flower collecting nectar or pollen.

She has long tongue which she pokes deep into the flower to suck up the sugar and water. It goes into her belly where there is a special honey stomach that stores it as it is carried it back to the hive. There she can bring it back up again, but this not vomiting, it has not been swallowed into the real stomach or gut. Honey is NOT bee vomit!

On a foraging trip a single bee may visit over 100 flowers depending on how much nectar she finds, but she will visit only one type of flower on each trip. Bees are colour sensitive, loving blue and purple, but they do not see red well. If you have purple lavender in your garden, just watch how bees love it!

A bee can carry about a quarter of her own weight, flying at about 25 kilometres an hour over a distance of several kilometres. She is often very tired when she gets back home.

Collecting pollen:

As she lands on the flower and buries herself in it she becomes dusted with the powdery pollen as you see in the picture below. Using her front legs (she has six, remember) she scrapes it backwards to her back legs which have special little baskets called corbiculae to carry it back to the hive. If you look at bees entering the hive after a foraging trip you will see the little blobs of colour on their back legs. Beekeepers like to see different colours of pollen because that tells them that the bees are visiting different flowers and getting a variety of this nutritious pollen food, just as we need a variety of food to remain well nourished.

Photographs by Elizabeth Frost

Pollen collection from a passion fruit flower. Note the dob of yellow pollen in the pollen baskets ("saddle bags") on the back legs of the bee in flight.

Photograph by Rod Bourke

Collected pollen is stored in honeycomb

Photograph supplied by Elizabeth Frost

THE BEE HIVE

A bee hive is actually the home bees live in but the term is used loosely to refer to the colony of bees within it. Remember that bees have been here for many millions of years, living in holes in trees, logs or even in the open.

A wild beehive built in the open. The sheets of honeycomb hang vertically with just enough space between them for the bees to move and work.

Photograph by Anthony Pyne

Wild bee hive high up on a pole in the open air

Photograph by Warren Burley

The bee hive as we know it is man's invention; a means of getting at the bees and the honey. Man has used all kinds of things as hives, but what is most associated in people's minds is this:

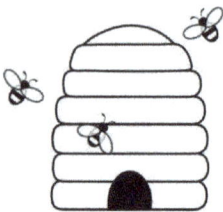

It is a straw hive called a "skep" and was used for many years long ago. It was fine for the bees but difficult for the beekeeper who had to damage the colony to get the honey from the honeycombs.

Some beekeepers have exploited the bees' use of hollow logs by modifying them to allow access. Beehive construction is limited only by man's imagination.

Photographs by Des Cannon

16

Over 250 years ago a minister of religion in America, Lorenzo Langstroth, invented the hive which is the basic type used today. It looks like this.

An empty hive

And this is what it looks like with bees in it

BEESWAX AND HONEYCOMB

What is beeswax? Wax is a fatty substance and beeswax is what the bees make to form an essential part of their homes and life-style. It is made from glands on their bodies, little droplets which they chew and mould into the form they need to make honeycomb.

This is an outside wild beehive with sheets of honeycomb showing the six-sided (hexagonal) cells, some open, some containing eggs, some capped over

Photograph by Anthony Pyne

Inside an empty hive showing wooden frames for the bees to build their honeycomb

Picture from internet free images

You may well wonder how these little engineers manage to make perfectly hexagonal honeycomb cells? They don't! The cells they make are round at first but physical forces pull them into six-sided hexagons. There is no wasted space in honeycomb.

Honeycomb is what the bees make with wax to store their honey and breed their young. It is a sheet of beeswax with many, many holes called cells. Each of these holes is six sided, called a hexagon. Shaped like this there is no wasted space. The bees store pollen and honey in some of these cells and in others the queen lays her eggs.

EVOLUTION OF BEES

Having evolved from the same stem over 100 million years ago, bees are closely related to wasps. Wasps are carnivorous, preying on other insects, while bees evolved as vegetarians. The oldest bee fossil identified is from the Eocene period, 40 million years ago. The defining characteristic of bees is collection of pollen as food for themselves and their developing young. Pollen for the bee is a protein and fat rich food; for the plant it is the male gamete which, like the sperm of animals, must be united with the female part of the flower to produce fruit and seeds which perpetuate the species. Thus bees evolved with flowering plants, called *angiosperms*, and they are interdependent, bees depending on pollen from flowers for food and flowering plants depending on bees for propagation by cross pollination.

While we are taught that modern man, *Homo sapiens*, developed over the last 200,000 years, the morphological structure of bees has not changed significantly in 30 million years. This tells us much about the efficiency of bee and its ability to survive.

BIOLOGY OF BEES

THE PLACE OF BEES IN THE ANIMAL KINGDOM…TAXONOMY

In the 18th century, about 1738, a Swedish biologist, Carl Linnaeus, published his work on a system of naming and classifying living creatures; the science of *taxonomy*.

Carl Linnaeus

Within this system the name of the honey bee is *Apis mellifera*. Creatures are each given its place, as below, where the honey bee position is indicated:

Kingdom: ……….. Animalia
  Phylum:………….Arthropodia (means jointed limbs)[1]
    Class: …………. Insecta
      Order: ………..Hymenoptera
        (means "membranous wings" and includes ants, bees, wasps)
        Family:……..Apidae (bees of various types)
          Genus:…. Apis (the genus is always given a capital initial letter)
            Species: mellifera (the species is always written in lower case)

The species name Linnaeus gave to the honey bee, "mellifera" means "carrier of honey" but bees carry nectar, pollen and water and **make** honey. Someone

---

[1] According to David Attenborough 80% of the animal life in the world are Arthropods, insects and arachnids (spiders and others).

after Linnaeus tried to change the species name to "mellificia" which means "maker of honey". But it did not stick and the name originally given by Linnaeus has stood the test of time if not accuracy.[2]

The genus *Apis* contains five species: our honey bee *mellifera*, giant bees *dorsata* and *laboriosa*, a dwarf bee *florea* and the Asian or Indian bee, *cerana*.

The species *mellifera* has several sub-species or races. Note that animals of different species do not naturally mate together and produce young. Those of the same species but different races can and do, and the resulting offspring is called a "hybrid". This is obvious if one thinks of the various human races, though we don't call mixed races "hybrids". The races within the species *mellifera* evolved in different geological areas, from southern Africa to northern Europe, and some of the names reflect their origin. They are:

mellifera mellifera: black German, English bees, the honey bee originally brought to Australia in 1822 in the convict ship *Isabella* (see later chapter "How was the European Bee brought to Australia?").

mellifera ligustica: Italian (Liguria is a coastal region of north-western Italy; its capital is Genoa.)

mellifera carnica: from the Austrian Alps

mellifera caucasia: From the Caucasus, the area south of Russia, north of Turkey and between the Black Sea and the Caspian Sea.

mellifera scutelata: the best known of five African races

Each of these races has special characteristics, some reflecting the climate of the country of origin. For instance, those from the cold areas, *carnica* and *caucasia* withstand the Australian winter better than *ligustica* from more temperate Italy.

---

[2] Learning the meaning of biological names of Latin or Greek origin adds to interest, understanding and memory.

The Italian bee is a lovely golden colour, like the queen bee in the centre of this picture. Other races are darker.

Photograph by Anthony Pyne

Hybrid queen bee

Photograph by Anthony Pyne

African bees have a sinister reputation for being very aggressive and have attracted the undeserved nick-name of "killer bees". Their stings are no more dangerous than those of other bees but aggressive bees deliver more of them. On the other hand, it is claimed that these African bees are more productive. In fact, they were imported into South America because of this characteristic. There they cross bred with local races of *mellifera* and the resulting hybrids are said to be even more aggressive and more deserving of the epithet, "killer bees".

Because of inter-breeding of races, most of Australian bees are hybrid but with a predominance of an individual race. Kangaroo Island is isolated from the mainland and boasts of its Italian bees.

ANATOMY

The anatomy of bees is similar to that of other insects with special configuration and features. The body is in three parts, head, thorax and abdomen.

The head is the seat of the senses; sight, hearing, smell and touch. Smell and hearing are mediated through the marvelous antennae, the long appendages waving about from each side of the head. There are five eyes; two large ones on either side and three small ones, called *ocelli*, on top of the head. (See heading "sight".)

The head contains the mouth parts including a tongue necessary for sucking up nectar from the depths of flowers. Some species have long tongues, while others have shorter ones limiting the variety of flowers they can forage on. Important glands in the head emit their pheromone odours to influence actions of the general population of the colony.

The bee brain, about the size of a sesame seed, contains one million neurons compared with 86 billion in the human brain.

The Thorax, consisting of three segments, is the organ of locomotion with a pair of legs on each segment and two pairs of wings on the posterior segment. The legs are jointed as implied in the name of the phylum , *arthropodia*.

Similarly the wings are finely membranous as implied in the name of the order, *hymenoptera*. Legs and wings are powered by strong muscles. (See later chapter "Flying ".)

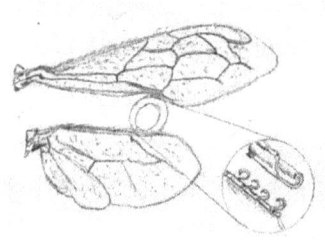

Bees have four wings but appear as one each side because they are locked together by fine hooks. They often separate when the bee is sick giving the dying bee a "K" appearance.

Drawing by Anne Creevey

The Abdomen consists of seven segments which can expand with the intake of 25-40 mg of nectar while foraging. It contains the alimentary system: oesophagus, honey crop, stomach, intestine and rectum.

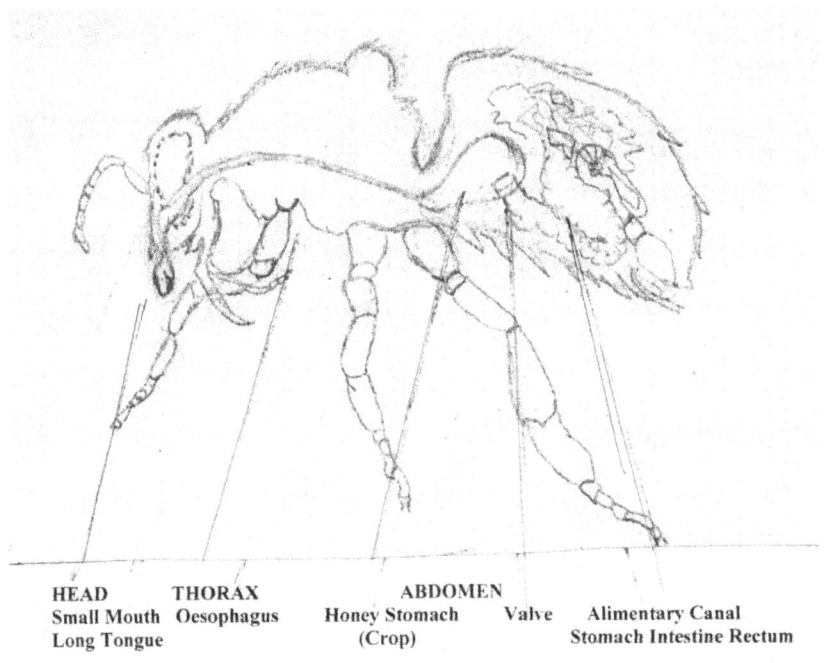

HEAD	THORAX	ABDOMEN		
Small Mouth	Oesophagus	Honey Stomach	Valve	Alimentary Canal
Long Tongue		(Crop)		Stomach Intestine Rectum

Drawing by Anne Creevey

The oesophagus leads into a crop, or honey stomach, which is where nectar is stored and carried back to the hive. It is separated from the rest of the alimentary tract by a valve, so when the bee regurgitates the nectar it is not really vomiting. Honey is not "bee vomit" as some disparaging cynics like to suggest.

In the hindmost part of the abdomen of the worker bee is the poison sac and sting. The sting is normally retracted inside and is extruded when the bee attacks. It is barbed and cannot be withdrawn from the skin after stinging a man or animal. The poor bee struggles and pulls till her bottom comes off leaving sting and muscular poison sac still pumping the venom into the victim, while she drifts away with entrails hanging out and dies. Queen bees have a sting with a difference. Theirs is not barbed, and for some reason she does not sting a beekeeper handling her. It is used to sting and kill rival queens, adult or in the unborn pupal stage. Drones do not have a sting. In its place are the genitalia, which, if used to fertilise a virgin queen, suffer the same fate as the worker's sting, coming apart from the poor fellow's backside to be left in the queen while he falls away and dies.

Cardiovascular, Respiratory and Nervous Systems are quite different from ours. There is a heart and aorta, but "blood", called haemolymph, bathes the tissues rather than being pumped around in blood vessels. It delivers nutrients to the tissues but not oxygen. Multiple breathing tubes along the body, called tracheae, provide oxygen and eliminate carbon dioxide by diffusion.

A small brain in the head is connected to multiple nerve centres along the body, called "ganglia" and these autonomously control muscular activity. A beheaded bee can still move its wings.

Glands and their products are of vital importance to the bee and the colony influencing nutrition, communication, defence and wax production. Wax glands are under the scales on the undersurface of the abdomen develop in middle aged bees whose duty is comb construction.

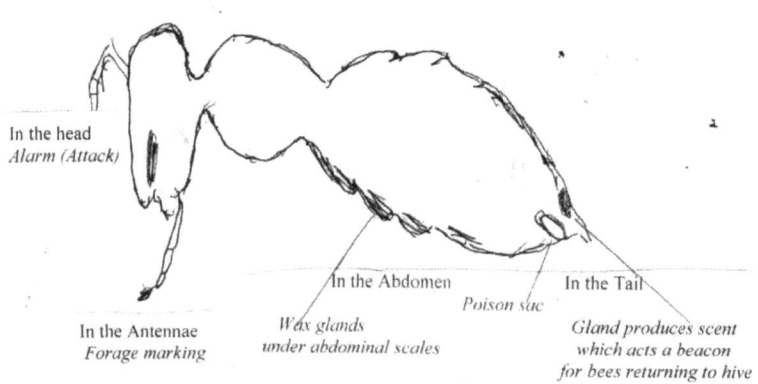

Glands of the bee

Drawing by Anne Creevey

REPRODUCTIVE BIOLOGY OF THE HONEY BEE (*Apis mellifera*)

We have seen that there are three casts of bees in the honey bee colony.

The queen is the only fertile female and her only responsibilities are egg laying and exudation pheromones which exert a vital controlling influence over the whole colony. In the wild a queen will live three or four years, but she loses fertility as she gets older. Hence, in colonies managed by beekeepers she will be replaced by a younger queen every year or two, (more below). A virgin queen emerging from her honey comb cell rests for a few days as her wings mature and then she embarks on mating flights. She flies out of the hive to an area where drones from other colonies fly in expectation of such an arrival and there they mate, on the wing, some distance up in the air. With several such flights the queen will mate with 15-30 drones, storing their sperm in her spermatheca for the rest of her life, up to three or four years. Within another few days she starts laying eggs, over 1,000 a day. She will never mate again and she will never fly again unless the colony swarms.

There are several biologically remarkable features about this mating process. Firstly, how she instinctively does not mate with drones from her own hive, her brothers. She flies well away from her hive to avoid doing so, though who knows if it might not occur occasionally by serendipity. Secondly, biological diversity resulting from multiple mating is very important for the health and strength of the colony. Just think of the consequence to this social colony if she mated but once with a drone of poor genetic quality, and there are species of bees in which the queen does mate only once. Scientific research using queens fertilised by artificial insemination (yes, this microscopic procedure is done regularly) has found that colonies resulting from a queen fertilised from one drone is weaker and less productive than those with queens mated with several males.

The worker bees, many thousands in a colony, are all female but infertile. They have the same genome as the queen, but they are different in body, in function and in fertility. We shall see below that this difference is dependent entirely on the nutrition of the larval stage of development. Worker bees work themselves to death, living about six weeks in spring and summer months, longer in winter when they are less active. Recent research has revealed that some bees, called "diutinus" bees, are destined to live longer. Worker bees warm the hive by generating heat with isometric muscular contraction, (contracting without movement) and recent research reveals that there are "heater bees" that do this more effectively.

The drone, the male bee, is remarkable physiologically. He has only half the number of chromosomes that the queen and worker bees have, a biological state called "haploid". He is the product of an unfertilised egg. He is nothing more than a flying sperm. His sole function in life is to fertilise a virgin queen and if he is successful in this enterprise he dies immediately.

There may be several hundred drones in a colony when it is breeding up in spring time, but none in winter. So where did they go? They live about three weeks in the breeding season but when this is over those no longer needed are disposed of by the workers, no longer fed and driven out of the hive.

Let us now look at the development of these different casts of the honey bee.

The worker. The queen lays a fertilised egg in a honeycomb cell which has been cleaned by young recently emerged workers and other young bees put a dollop of royal jelly on it. Royal jelly is a high protein food made up in the bodies of worker bees from pollen and honey. With good light the small, white cylindrical egg can be seen in the bottom of the cell leaning on the wall, one egg per cell. In three days it hatches into a larva, a tiny white grub the size of the letter "c" in newsprint. Fed intensively by nurse bees with a little royal jelly initially followed by a mixture of honey and pollen products, called "bee bread", it grows rapidly for the next six days to fill the cell completely.

Eggs (right of picture) and larvae (on the left)

Picture from Wikipedia free images

Well fed larvae grow rapidly, filling the cell in six days

Picture from Wikipedia free images

It is then capped over with wax by another team of young bees. Now called a pupa, it undergoes a process called "metamorphosis" which is one of the unexplained miracles in biology. This white grub no longer fed, moults (sheds its coat) several times and in 12 days changes into a brown adult bee which slowly chews the lid off her cell and emerges into the frenetic hurly-burly of the active beehive. Then, another miracle, this little hairy fluffy bee goes to work right away and with no schooling or tuition and continues to perform a range of sequential functions throughout her few weeks of life.

Worker bee emerging from honeycomb cell, 21 days after the egg was laid

Photograph supplied by Elizabeth Frost, DPI

Now think of this timetable of development and life, three days for the egg to hatch, six days as a larva and twelve pupating, total 21 days egg to adult, and then living six weeks as a mature adult. If man, *Homo sapiens*, followed the same timetable, we would live 18 months!

Another concept to ponder over: how nature makes and destroys so relentlessly. So much involved in producing this bee, and then destroying such a marvelous creation by death after such a short time.

While so pondering, think of the leaves and flowers. If you have seen a leaf under a microscope you will agree that it is an architectural miracle. It is a complicated miniature power generation unit harvesting energy from the sun. And a flower has a beauty nothing made by man can match. Yet nature ruthlessly discards these marvelous creations in millions!

The Queen

We have seen that the queen has the same genome as a worker bee, yet her body and function are so different. So how does a queen become a queen? The bees actually make a new queen when the old one is aging and failing, or has been killed, perhaps by a beekeeper accidentally squashing her. Or in contrast when the colony is preparing to swarm because it has so increased in numbers that there is no longer adequate space in the hive. The falling level of queen pheromone pervading the hive supplies the trigger for an instinctive corporate decision to make a new queen from one of the old queen's eggs. If the old queen is dead and there is none of her eggs or very young larvae, the colony is in real trouble, (more below).

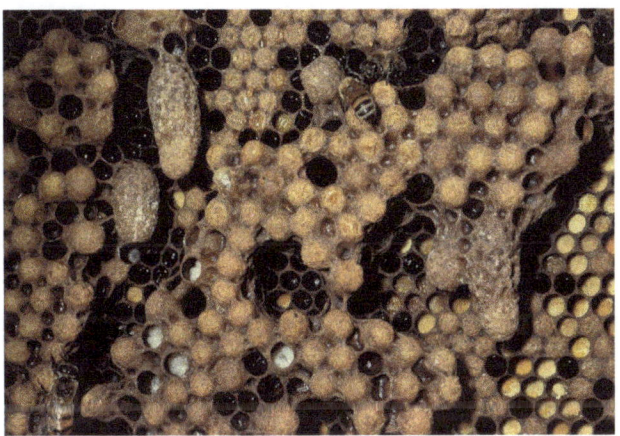

Multiple queen cells on comb with yellow and orange pollen and some white larvae.

Photograph supplied by DPI

Selecting an egg, often more than one, the bees build a large cell around it and feed it with extra royal jelly throughout its larval period. The queen cell is quite distinctive, mamilliform[3] in shape, always hanging down from the comb.

---

3 Mamilliform: shaped like breast

Three queen cells on a frame of comb with worker brood, two below the bar and one on the other side

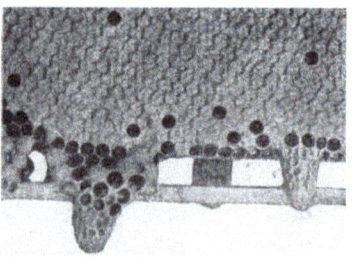

Picture from "Bees and Honey" WA Goodacre, "Bees and Honey", NSW Department of Agriculture, 1941

This photograph shows multiple queen cells hanging downwards from the comb, a drone cell with the dome shaped cap, and some worker cells with only slightly raised caps

Photograph by Doug Somerville, DPI Technical Officer

When preparing to swarm multiple queen cells are built, usually along the bottom of a comb. They are timed to emerge just after the old queen and about half the bees have flown out to find a new home. The first of these new queens to emerge will sting the others to death through the wax cell, and if two emerge together they will fight to the death. Too bad if one is dead and the other damaged; perhaps that's why the bees in their wisdom often make three or more.

Traditional teaching is that it's the extra rich feeding with high protein royal jelly which directs the egg and developing larva on the path to being queen rather than a humble worker, but new research evidence suggests the opposite. It is perhaps the vegetable product flavonoids contained in the bee bread fed to the worker larvae which prevents their sexual development. One of these, p-coumaric acid, has been shown to suppress ovarian development. This theory seems very likely as we know the enormous influence maternal diet during pregnancy has on human embryonic and foetal development.

SWARMING

Have you seen bees in a ball, like this, hanging from a tree?

Photograph by Tony Mulquiney

Or swirling in the air like a cloud?

It is called "swarming" and it happens in spring and summer when bees are leaving home. When you grow up you will leave your parents' home, moving out to a place of your own. That is what bees are doing when they swarm. In the warmer months of spring and early summer there are lots of flowers about and with plenty of food the bees multiply with numbers growing and growing till there's not enough room in the hive for so many. So bees do what families do as children grow up, some move out. They very cleverly breed a new queen to take over the hive so the old queen can take off with half the worker bees to find a new home.

On a warm sunny morning they say "goodbye" and tumble out of the hive, taking to the air and swirling around in a big cloud. Moving to a nearby tree they settle in a big bunch, like a football. There they rest while they send out scouts to find a suitable new nesting place. When the scouts return with good news they all take off again to fly to the new home. Unfortunately that home may be in *your* home, the wall or the roof.

Meantime, back at the old hive, the new queen gets on with breeding new bees there. There are now two colonies, and that's how nature increases the number of bee swarms or colonies.

If you happen to come across swarming bees like this, don't be afraid. They are not interested in attacking you; they are too busy finding a home. But don't go near them, don't throw things at them, don't spray them with water. Get someone to call a beekeeper who will catch them and give them a good home well out of your way.

Swarming bees are not interested in stinging

Photograph from "Bees and Honey" 1941 by WA Goodacre

Footrot Flats strips reproduced by permission of Diogenes Designs Ltd

SOCIOLOGY OF BEES, Who controls the colony and how?

Here is another miraculous mystery of nature to ponder over. Who, or which bee, or bees, make the corporate decisions leading to actions involving the whole swarm? Popular concepts would have it that the queen rules the swarm, and so she does by means of her smell pheromones. But when the level of this influence falls as it does when she gets older, or disappears when she dies, which bee or bees make the decision and take the initiative to breed a new one? Recent research suggests that the decision of an overcrowded colony to swarm is made by the young nurse bees who feed the brood. But we don't know how.

So let's look a little more at these controlling pheromones. A hormone is a chemical substance produced within the body to influence and control the function of other parts or organs of the same body. For example, in humans insulin produced by the pancreas controls the metabolism of sugar by the rest of the body, growth hormone from the pituitary gland at the base of the brain influences the growth of the skeleton and the body generally. A pheromone is a chemical produced in the body of one individual to act outside that body on others and direct their activity. These pheromones are produced by glands, bunches of specialised cells in various anatomical sites. Worker bees also have pheromone exuding glands. Most interesting are the Nasonov glands in the hindmost part of the abdomen. A pheromone emitted from this gland by the bees at the hive entrance acts as a beacon to guide returning bees home. An alarm pheromone from the stinger glands alert adjacent bees to danger and one from mandibular glands gives the signal to attack. A beekeeper knows that a stinging bee leaves traces of this substance on the sting site, attracting other aggressive bees to add more stings, so as well as flicking the sting out he will blow smoke on the site in the expectation (perhaps more hope than expectation!) that the smell of smoke will overcome that of the attack pheromone.

Of all the pheromones pervading the colony, those produced by the queen are most important.

Queen with attendant bees grouped around her, grooming her and bathing in her pheromones

Photograph by Des Cannon

This often reproduced photo of the queen shows her surrounded by worker bees radially disposed around her. They are grooming her, cleaning her, and at the same time they are becoming infused with her pheromone smells and as they subsequently move around the hive in contact with other bees on their regular duties they spread it far and wide. As indicated previously, while that smell is there, the colony will function as an organised community. That hormone also maintains the suppression of the latent ovaries of the female workers, suppressing their egg production. In a situation where the queen is lost and there is none of her eggs to make another queen, one or two workers now released from the ovarian suppressing queen hormone may develop ovarian activity and lay eggs. But laying workers are unmated and the eggs will produce only drones, bees with half the complement of chromosomes. That colony will die and a beekeeper cannot help. He can't find the laying worker because she looks the same as thousands of others and if he introduces a new queen the worker who fancies herself as a queen may kill her.

PHYSIOLOGICAL FUNCTIONS

Nutrition

They need the same types of foods as we do: a varied diet consisting of sugars and some fat for energy, protein for building bodies, vitamins, minerals like salt, and water. But they are vegetarian. Unlike their cousins, wasps which eat other insects, bees eat only food from flowers, nectar from which they make honey, and pollen.

Water is needed not just to drink as we do, but the bees also use it in their hives to keep them cool in summer.

Sight

Bees have five eyes, two large ones on the front of the head and three small ones, called *ocelli*, on top of the head. Their functions differ: the *ocelli* detect light intensity, the large ones are for main vision. They detect colour, more at the blue end of the spectrum than the red and they are sensitive to ultraviolet light. Recent Australian research has cast doubt on traditional concepts of red blindness claiming that their vision is basically different from ours. They can see where they are flying even on a cloudy day because they are sensitive to ultraviolet light. It is completely dark inside the hive, but somehow they see to move around and do their work. When you step out of a dark room into bright sunlight you squiz up your eyes till they adjust to the light, but bees taken out of a dark hive into strong sunlight do not seem to notice. They don't have eyelids so they can't close their eyes as you do.

Today's fact
Bees can see ultra-violet light, but can't see red.

Colours are important to bees. If you look at bees on the flowers in the garden you may notice that bees love lavender, purple flowers. They love blues and purples because they can see them best, but they are not good at seeing red. So if you want to attract bees to your garden, plant purple flowers, Lavender, Rosemary, Basil.

Hearing

Insects don't have ears like ours, but they can hear some sounds. A cricket's "ears" are in its knees! A bee's "ears" are in the antennae, the two worm-like feeler things sticking out from each side of the head. They are the seat of many of a bee's senses, including hearing and smell. A bee hears low sounds but not high ones.

Smell

Smell is very important for bees. In fact, their whole life is governed by smells of many different kinds. They certainly smell honey. If you leave honey outside in the open, bees will soon find it. If only one bee finds it, she will taste it, fly back to her hive, tell her mates about it and soon there will be lots of others at the honey pot. As well as colour, flowers attract bees by the smell of nectar in them.

Like hearing, the organ of smell is in the antennae, and since there are two of them, bees can tell the direction the smell is coming from. Can you do that?

Antennae waving out from the front of honey bee's head

Painting by Gina Cranson

Bees at the entrance to a hive also give out a smell which acts as a guide to bees returning from a foraging trip, just like a radio beacon guiding an aeroplane back to the airport. Another important smell is one that gives the signal to other bees to attack and sting. When a bee attacks an intruder to the hive, like a beekeeper opening it to take honey, she gives off a smell which tells other bees to get into the fight with her. And if she stings the beekeeper, that smell is left on the site and other bees attack the same site.

Taste

Taste is closely related to smell. So it should not surprise you to know that the bee's sense of taste is also located in the antennae. When a scouting bee finds a supply of nectar she will taste it to see if it is good and take some back to the hive for others to taste.

Antennae waving out from the front of a native stingless bee's head

Painting by Gina Cranson

Walking

One might think their six legs would trip over each other when walking, but they don't! Bees walk surprisingly quickly and legs are also used for more than walking. The back legs have little pollen baskets called "corbiculae" in which pollen is carried back to the hive. When full these are clearly visible and beekeepers like to see lots of such loaded bees entering the hive. It is an indication that the colony is breeding well, and different coloured pollens confirm that the wide range of proteins is in their diet. Front legs are used to scrape the pollen from the flower to the corbiculi. They are also used to do the cleaning and packing in the hive.

Flying

Flies have two wings; bees have four but it looks like two because the wings on each side are locked together by little hooks. They are very fine and delicate and eventually wear out. Wings vibrate at over 200 cycles per second, allegedly faster than nerve impulses, so they resonate, or beat more than once in response to each nerve impulse. They propel the bee through the air at about 24 km/hr and carry her about 800 kilometres in a lifetime, eventually becoming tattered and torn, leaving the bee unable to get home and to die. It has been estimated that if loaded with honey a bee could fly 60 km before running out of fuel.

Who do you think would win in a race between a pigeon and a bee? Well, such a race has been staged and the bee won by a nose. The marked bee was entering her home hive while the pigeon was circling the loft above.

How far do you think a bee will fly to gather nectar or pollen? The honey bee will go as far as three, four or even five kilometres if she has to, but she will be very tired after such a long flight and obviously prefers to forage close to home.

Navigation

Ever wondered how bees find their way to that patch of clover and then unerringly wing their way back home? How do they navigate?

They have several methods.

First, they steer by the position of the sun. They can do this even on a cloudy day because they are sensitive to ultraviolet rays which penetrate clouds. Though the sun is constantly shifting its position across the sky, they can compensate for this, and even pass this information on to other bees in the hive as they recruit them to get out there and forage. Polarity of sunlight also has an influence.

Bees are so sensitive to magnetic fields that a strong magnet can distort the way they build hives.

A second method is visual. They note the landmarks on their route; trees, buildings, roads. Bees' perception of colour is important in identifying flowers, sensitive to the lower frequencies like blue and purple but less to the higher red range. This use of visual navigation is very important for beekeepers to be aware of when moving hives. They know that they must move them only a metre or so at a time or move them several kilometres away. If moved a very short distance, a metre or so, the returning bees can still find the hive using their previous orientation guides, but if shifted more than a metre with no major change in surrounding visual landmarks they will return to the old site. When moved longer distances with completely different landmarks they must to reorientate and re-programme their navigation. A third method, less well understood, is the use of the earth's magnetic field, sensitivity to its influence made possible by the presence of minute amounts of iron compounds (magnetite) in the abdominal region. Just how this works is unknown, but research has revealed that cancellation of the magnetic field around the hive interferes with their orientation and communication.

Finally, the bee's acute sense of smell is also an aid to navigation as they approach home on the return journey. A pheromone emitted by the guard bees from a gland in their tails effectively acts as a homing beacon. The same pheromone is left at productive foraging sites to guide other bees to it.

Communication, do bees talk to each other?

'TYPICAL, THE ROSE FARM MOVES AND YOU BRING AN OLD MAP!'

Beekeepers know that exposing honey outdoors in the vicinity of hives soon attracts bees in ever increasing numbers, eager to lap up this easy source of food and take take it back to the nest. How do they find it?

It takes just one bee scouting or cruising by to smell it, sample its taste and hurry back home with the news. There she quickly tells her mates all about it, the direction and distance from the hive and the richness of the food source. That this one bee can do this was demonstrated experimentally in the early 1900s in Europe by Maurice Maeterlinck. He captured and marked a bee foraging at a honey bait and then allowed her to fly home to the nearby hive. There he waited for her to emerge, captured her again and prevented her leading other bees to the honey. But they still found it!

So how did she tell all the other bees about it?

This was extensively researched later by another European, Karl von Frisch, whose results were published as recently as the 1967 in a book called "Dance Language and Orientation of Bees". Using a glass observation hive he was able to observe their communication methods and found that though it is dark inside the hive they communicate visually by dances. They give explicit information about the direction, distance and food value of the source. In addition to these visual methods, the passing of samples from bee to bee gives information by smell and taste. Von Frisch observed, described and interpreted some of these dances in detail. He actually learned bee language!

Using one he called the "Round Dance" a returning bee tells of a food source a short distance away, less than 15 metres. She turns in small circles, reversing frequently and passing samples of the food to observing recruits, the vigour of the dance reflecting the richness of the source. This gives no indication of direction; the recruits having to find it themselves by circling, aided perhaps by a pheromone smell left at the site by the first bee.

A second, the "Waggle Dance", tells of a source further away, and the instructions are more specific, indicating direction, distance and the richness of the find. The bee runs quickly forward on the comb for a short distance, shaking her body vigorously from side to side making a buzzing sound, then turning to a particular angle to the vertical (remember she is on the side of a vertical comb). This angle indicates the direction of the source in relation to the position of the sun, while the length of the run, over several honeycomb cells, gives the distance.

These communication techniques are obviously imprecise, but in one experiment 89% of bees present at a briefing found a close food source in five minutes. For longer distances, a directional error of 9-12 degrees has been estimated, and distance error of 2-10%.

Still very clever! I wish I was as accurate in interpreting directions my friends give me.

DISEASES, PARASITES AND PESTS

Do bees get sick?

They do! **But a sick bee cannot make you sick, and if you are sick you cannot give your illness to bees.** Bees don't get a cold, and you don't get the same illnesses that bees get. Bees can be attacked by germs, bacteria and viruses, and there are parasites (other insects and creatures) which make big trouble for bees.

Beetles, moths and their grubs like to live in beehives and cause a lot of damage, even killing a colony. They add a great deal to the work of beekeepers.

Lots of other things also bother bees. In Queensland beehives must be kept high off the ground because cane toads love to sit on front of a hive and eat the bees as they enter and leave. There are birds called Rainbow Bee-Eaters which also do that! Some moths love to lay eggs in hives of weak bees and their larvae make a mess of the honeycombs.

Here let us focus in more detail on some of these problems. Let it be repeated that bees do not infect humans with their germs and humans do not pass theirs on to bees. You can't give a bee a cold!

Over 70 years ago when this writer commenced beekeeping there were diseases and the descriptions given by the books printed then are still relevant today. But those diseases, still with us, are more of a problem now and there are more of them to make trouble for beekeepers.

(What is the difference between a disease and a parasite? Not much. We think of disease as a sickness caused by invisible organisms such as bacteria, viruses, fungi, by poisons or malnutrition, while the term "parasite" makes us think of other living creatures, larger than bacteria, which are detrimental to the host by attacking it or interfering with its life giving activities. Some parasites are actually beneficial to the host, a relationship called "symbiosis", literally meaning "life together ".)

There are diseases and pests we have in Australia, and some we don't have (yet).

Diseases and pests we have in Australia

The most important bacterial disease in this country and worldwide is **American foul brood** caused by a spore bearing germ called *Paenibacillus larvae* which infects developing larvae and pupae and kills them. It is resistant to antibiotics, incurable, highly contagious, and its heat resistant spores can survive for 30 years on equipment. It is a fatal disease, killing the affected colony, and in the process it is likely to infect others. It is spread by bees and beekeepers. Bees spread it by entering the infected and weakened hive, to steal its honey. They become contaminated with bacteria and spores which they then take back to their own hives. Beekeepers spread it by contaminating their hands and tools as they move from hive to hive in their apiaries, or by leaving honeycomb or honey outside an infected hive for bees from other hives to forage on and become contaminated.

Beekeepers must learn to recognise this disease and deal with it according to laws governing beekeeping, amateur or professional. These demand that the bees in infected colonies be killed and all hives and equipment used on them be burned or irradiated. Only fire or irradiation kills the spores of this deadly bacterium.

Other less lethal but serious diseases are encountered frequently. One called European foul brood is also a bacterial infection of developing larvae and pupae. It is similar to American foul brood but the organism is non-spore bearing and is sensitive to antibiotics. If the beekeeper is in doubt about identifying these diseases he has recourse to a laboratory for scientific confirmation.

There is a commonly encountered infection of the gut of the adult bee by a single celled microsporidium organisms called *Nosema apis* and *Nosema ceranae*. It causes diarrhoea and malnourishment. There are no specific signs of this *Nosema* infection, the colony just becomes weaker and weaker and the beekeeper must be alert to the possibility and take steps to get a laboratory diagnosis. It is very difficult to treat and control.

The small hive beetle (*Aethena tumida*)

This pest came to Australia in 2002 from South Africa. First detected in Western Sydney, it has spread widely and wreaked a havoc of destruction.

The little black beetle feeds on rotting fruit but especially loves beehives. Its eggs laid in the honeycombs hatch into larvae which feed there, make a mess, then exit to pupate in the ground nearby. Emerging adult beetles re-enter the hive and can fly long distances to other hives.

Small hive beetles in a hive with honey bees. Wikimedia Commons.

Adult small hive beetles

Photograph supplied by DPI

It is not the adult beetle which does the damage, but its larvae.

Photograph supplied by DPI

Eggs, hundreds of them, are laid in the hive and on the combs and hatch into larvae, off-white grubs, which feed on pollen, honey and brood. They leave slime over the combs and cause fermentation of the nectar and honey, rendering it useless. Bees recognise the beetles as invaders and though they can't sting them because of their tough carapace, they do harass them and try to drive them out. But with their rapid reproduction beetles can destroy even a strong colony.

Multiple small hive beetle larvae sliming and destroying honeycomb

Photograph supplied by DPI

Control is difficult. They are insects, but so are bees. Not only are bees sensitive to insecticides, but chemicals which could contaminate honey cannot be used in the hive. Traps, inside and outside the hive, have been designed to capture them, their variety limited only by the imagination of researchers and beekeepers. It is a matter of control rather than elimination and extensive research is focused on biological methods.

This is an "Apithor", an internal device called a harbourage trap. Inside it are insecticide impregnated tunnels so narrow that the bees can chase the beetles in but cannot go in themselves.

This is a "Beetltra" trap which fits under the hive with lime or diatomaceous earth[4] in the tray. There are slots in the bottom of the hive and the bees drive the beetles down to their death by desiccation in the tray.

There are many other such traps for use inside, outside and under the hive.

This pest has made beekeeping, commercial and amateur, much more difficult and has added significantly to the cost of producing honey.

Diseases and pests we don't have (yet)

Varroa destructor

The most devastating of these exotic pests is a mite called *Varroa destructor*. A mite or *Arachnid* is like a miniature spider, the size of a pinhead. As its taxonomic name implies, it has been a destroyer of bees all over the world---except Australia. At the time of writing this is the only continent free of it, but it is only a matter of time before modern travel and trade practices bring it to the country. The mite attaches itself to the body of the adult bee, penetrates its integument and sucks its haemolymph, (life "blood"), weakens it and leaves a wound vulnerable to infection by viruses. It is sensitive to insecticide, but so are bees and its management and control are a challenge for the beekeeping industry. It is a major threat to bees and beekeeping. Surveillance hives strategically placed near all Australian ports of entry are monitored to detect the pest when it eventually arrives.

---

4 Diatomaceous earth is sedimentary rock in fine particles. It is used in swimming pool filters.

Colony Collapse Disorder

News media in recent years have made much of losses of bees in the northern hemisphere. With bitterly cold winters they always lose some hives, but losses have increased to an alarming degree. As a result of research seeking an explanation a new condition has been described and called "Colony Collapse Disorder". It is not a disease entity, but a syndrome[5], a consistent concurrence of several features characteristic of the condition.

Originally called "hive dwindle", it was first noticed that colonies were failing with decreasing population numbers. There was an active queen and lots of young bees but no oldies. There was little food storage and remembering that the older bees are the ones that leave the hive to collect nectar and pollen it was apparent that they were not returning from their foraging trips. Some years of observation, thought and research have revealed multiple causes. First is the widespread use of insecticides in modern agriculture. Most extensively used are a group called "neonicotinoids", "neonics" for short. Nicotine is a substance in all of us in minute quantities as a neurotransmitter; it is involved in firing off nerve endings and transmitting signals, for example to muscles to make them move. In massive concentrations it is a deadly poison. In agriculture it is called a "systemic" insecticide. Soaking seeds with it results in its pervading the whole plant including the nectar and pollen. Consequently the foraging bees get a dose of the poison, not necessarily enough to kill outright but enough to cause disability, interfere with navigation, delay return to the hive and perhaps prevent bees getting home. Further, if they do make it home, their immune systems may be so compromised by the poison that they succumb to other diseases that would otherwise be resisted. Hence the disappearance of older bees and the poor nutrition of the colony.

A recent article by a thoughtful scientific beekeeper would add the stress to bees caused by modern commercial (and perhaps some amateur) beekeeping techniques. If you see a documentary called "More Than Honey" you will

---

5 Syndrome: A set of symptoms occurring together to cause or constitute a recognisable condition; not a specific disease or entity of itself

agree with a beekeeping colleague of mine who described some American commercial practices depicted as "brutal".

If you add the scourge of northern hemisphere beekeeping, the Varroa mite, to all these factors, you have the picture of Colony Collapse Disorder.

Does it occur in Australia? At least one researcher claims that it has always existed in minor form, but not to the extent found in the northern hemisphere. With agricultural practices moving more to huge monocrops and the necessary use of insecticides Australia may well beware.

Consider the imponderable. Can we produce enough food for the world's millions *without* insecticides? Can we pollinate our crops adequately *with* insecticides endangering the bees and other pollinating insects? If our major food crops are not properly pollinated how would we be fed? Neonicotinoid insecticides are widespread in the environment, recent studies in Switzerland revealing that 75% of global honey is contaminated and in North America 74% of rivers draining into the Great Lakes tested positive for these pesticides. What is that doing to aquatic life? A wise man (some say it was Albert Einstein) claimed that if bees are lost mankind would not survive more than a few years.

THE ASIAN BEE, A*pis cerana*

This bee is a pest. Some years ago there was grave concern about the arrival in Cairns (north Queensland) of a swarm of Asian bees which had hitched a ride on the mast of a ship. There were good reasons for concern. The bee looks like a small edition of the European honey bee, but has many undesirable features. Having evolved in the tropics it has never had any need to store honey for winter, so it doesn't. It just collects enough for day to day consumption, like living hand to mouth, no use to us for honey production.

Multiplying rapidly it has a strong propensity to swarm and so invade houses and trees in urban environments, and it competes for food with our established European bee.

But worse, much worse, it carries the dreaded Varroa mite to which it is resistant. Hence the concern when it arrived and the efforts by Queensland Biosecurity staff aided by professional and amateur beekeepers to eliminate it. But the attempt to destroy it failed, and efforts to contain it have not prevented its spread south. Mercifully, however, it did not bring Varroa with it.

Left: Asian Bee, Apis cerana   Right: Honey, Bee Apis mellifera

Picture supplied by Queensland Department of Agriculture and Fisheries

BEE STINGS

Have you been stung by a bee? Hurts, doesn't it? People often talk about being "bitten" by a bee, but bees don't bite, the sting is in their tail.

> A bee's front end is sweet and kind,
>
> But never trust a bee's behind.
>
> A bee will sting if it can sit,
>
> So always stay in front of it.

Only the worker bees sting. The male drone cannot sting; the queen can but she never stings us, and if she did she would be able to pull the sting out because her sting is not barbed. She uses her sting to kill other rival queens for there can be only one queen in a colony. The worker bee's barbed stinger is normally inside, so if you are looking at a bee you will not see it. When she attacks, the stinger comes out and drives its way into your skin. Because it is barbed like a fish hook the bee cannot pull it out; she pulls and pulls till her bottom comes off, leaving the stinger with its poison sac in the skin. The poison sac has muscles which go on pumping the poisonous venom into the victim, so **you must get the sting out immediately.** Do not pinch it out, that squeezes more poison into you, **flick it out with your finger nail.** The longer the sting stays there the more it hurts, the longer the pain lasts and the worse the reaction.

What happens to the bee? She is like a suicide bomber, she dies. She can be seen flying weakly away with her insides hanging out from her bottom. Poor bee! She is worse off than you; she gave her life in the defence of her colony.

Bee sting on a glove. See the poison sac which continues to pump venom after the bee has gone

Photograph by Dean Morris, supplied by Elizabeth Frost

Bee sting on finger

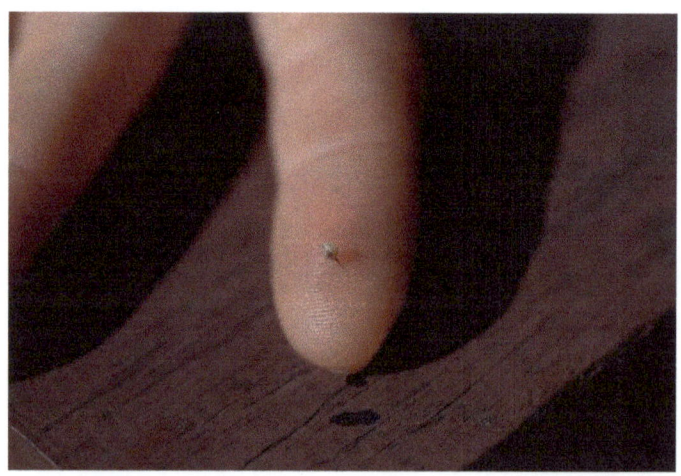

Photograph by Dean Morris supplied by Elizabeth Frost

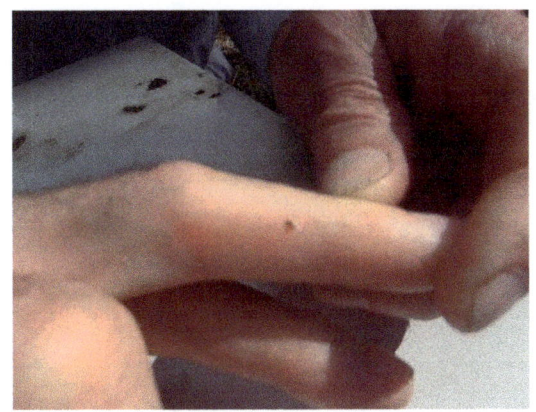
A bee sting on the author's finger. Removal was delayed taking the photograph, so the finger hurt for a few hours and swelled for a few days

What does the bee inject into you? A tiny, tiny droplet of bee venom, a poison. Though it is a very strong poison it will not kill you unless you are allergic to it. The part will swell, and if stung on the hand the swelling may extend up the arm, go red and get itchy. It may take many hours or several days to get better. Some people like to put honey on it and that helps. Medicinal ointment or cream containing the steroid cortisone is the best. If a person's body system of immunity is healthy, future stings cause less and less reaction, fortunately for beekeepers! But don't let any seasoned beekeepers tell you it doesn't hurt, it still does, but they are used to it!

Unfortunately some people have body immune systems which go wrong, and their reactions to repeated stings get worse and worse. If stung on the hand they swell up on the hand, arm and even the face. They may feel faint and have difficulty breathing. This dangerous response is called an allergic reaction. "Allergy" means "other reaction", an abnormal and unwanted response to a stimulus. In people who are allergic to something such a sting, the offending agent provokes the release in the body of a substance called "histamine" which causes the unpleasant and dangerous features of allergy; local swelling, redness, itch, remote swelling, difficulty breathing, fall in blood pressure and in the most severe case, death. This extreme allergic response is called "anaphylaxis" and demands the immediate administration by injection of adrenaline. In Australia about one person per year in dies of beestings.[6]

---

[6] The Australian Bureau of Statistics records that from 1960 to 1981 25 people died "shortly after a bee sting", whether directly or indirectly due to it.

"The first recorded fatality from stinging insect hypersensitivity probably occurred in Egypt in 2621 BC. The victim was King Menes, the first king of the first Dynasty, founder of the city of Memphis, and diverter of the Nile. An account of his death is found on the walls of his tomb. There is some ambiguity in the translation of the hieroglyphics as the symbol for a wasp is very similar to that for a hippopotamus. However, as no hippopotamus stings have been recorded since that time, the former translation seems secure!"

From a medical paper published by Dr David Sutherland, Newcastle Immunologist.

People who know they are so allergic carry a self- injector called an "Epipen". This is a device to self- inject adrenaline. No skill is required, just the knowledge of how to use it.

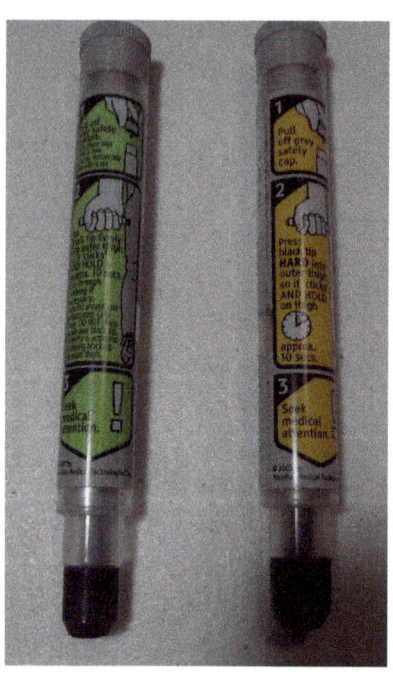

How to give EpiPen®
adrenaline (epinephrine) autoinjectors

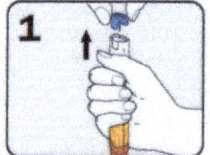

1. Form fist around EpiPen® and PULL OFF BLUE SAFETY RELEASE

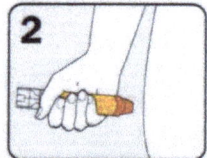

2. Hold leg still and PLACE ORANGE END against outer mid-thigh (with or without clothing)

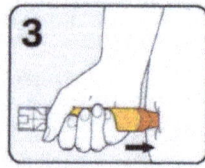

3. PUSH DOWN HARD until a click is heard or felt and hold for 3 seconds REMOVE EpiPen®

DO NOT PUT FINGERS OVER THE ENDS OF THE INSTRUMENT. GRASP IN A FIST AS SHOWN

Administering adrenaline using the Epipen is first aid for anaphylaxis, just the beginning of care. Further medical care is needed urgently. Doctors would also administer steroid (cortisone) medication and monitor progress in an intensive care environment.

Allergies can be treated by a desensitisation process which involves giving at first a very tiny amount of the offending agent, called an "allergen", and slowly increasing the dose as the immune system learns to cope with it. The process may take a year or more, sometimes for the rest of the person's life, and requires the service of a specialist doctor called an immunologist.

Many people claim to be "allergic" because they swell up at the site of the sting or on the affected limb. This is not necessarily allergy, just the direct effect of the tiny dose of a very potent poison. As indicated previously, it is very important to flick the sting out of the skin as quickly as possible as the muscular poison sac remains attached and continues to pump venom for a few seconds. This story illustrates:

In a hands-on teaching class for beginners in beekeeping I was not wearing gloves and was stung on the left hand. Instinctively I flicked the sting out immediately, and then thought, "That was a pity, a missed opportunity, these eager beginners have probably never seen a sting close up." So when stung on the right hand a little later I delayed removal of the sting to show the poor bee struggling to pull the barbed sting out of my skin, eventually pulling her bottom off and, with trailing entrails, falling away to die leaving the sting and poison sac pumping away in my hand. That took time, not much but enough. The left hand hurt briefly, for when busy with attention focused on working a hive such a sting is more of a nuisance than a distraction. The right hand really hurt and was still hurting when I drove out from the apiary a few hours later; it swelled and remained red and swollen for a day or so and itched for most of the week. **So flick that sting out immediately!**

And what can you put on a bee sting site to avoid this distress? Some say put honey on it, or some other patent remedy, but what really works is a steroid (cortisone) cream or ointment. Another story illustrates how effective it is:

On another occasion I foolishly approached an open hive without a veil and received multiple stings around the face, neck and ears. (Richly deserved. You can't see to remove stings around the face, they hurt and are dangerous near the eyes, so always wear a veil no matter how "brave" you feel.) Anticipating that my face would be swollen beyond recognition next day, I generously applied a strong steroid ointment as soon as possible and repeated it later in the day. My face was no longer hurting as I went to bed and the mirror reflected a visage with no evident swelling. And so it remained next day.

BEEKEEPERS AND BEEKEEPING

While modern man has existed for some 200,000 years, bees evolved with flowering plants several million years ago, the oldest fossil dated as 100 million years old. No doubt man and his forebears have enjoyed honey as a food from time immemorial. Australian aborigines certainly ate the honey produced by the Australian stingless bee. It is not known how long man has cultivated bees. Beeswax is long lasting and it is used as a marker for identifying bee products but does not tell whether fragments of pottery so identified came from a beehive or a vessel storing collected wild honey and comb. Honey, tasted and apparently still edible, has been found in Egyptian tombs of some 5,000 years ago. References to beekeeping can be found in the writings of Aristotle (344-322 BC) and Virgil (70-19 BC, the Roman Augustan period).

During this long history bees have changed little if at all, but beehives have developed from pottery tubes or vessels and later straw skep hives, (still the best known symbol of beekeeping), to the sophisticated technology of the newly invented Flow Hive. It is important to realise that beehives, where domesticated bees are housed, are entirely for the convenience and access of the beekeeper; bees have thrived happily in hollow logs and cavities for millennia.

People who keep bees are called beekeepers or "apiarists" and where they keep their hives is called an "apiary" because the biological generic name of the honey bee is "Apis". Do beekeepers get stung? Of course they do, but in most healthy people reactions get less serious as they get older. We have already learned the importance of removing the sting quickly. Experienced beekeepers with a tolerance to stings may not wear full protective suits or gloves while working with their bees, but they will wear veils to protect faces and eyes. Stings around the eyes are dangerous, stings on the face hurt and you can't see to remove the sting.

Well protected beekeepers working on a hive

Beekeepers with veils, suits and gloves. But the woman on the right has just been stung through her clothing.

Beekeepers have another trick. They use smoke on the bees to calm them and make them less inclined to attack.

A smoker is just a barrel containing fuel like pine needles, leaves or old bags lit to produce smoke which is blown out of the barrel with air pumping bellows. How does smoke calm the bees? You might read or be told that they think there is a fire and rush into the hive to fill themselves with honey ready to get out and fly away, but this isn't true. The smell of the smoke overwhelms the smell and of the attack pheromone and diminishes their urge to attack.

BEEHIVES

Humans have been keeping bees for thousands of years. Scientists called "archeologists" have found old beehives from long ago, some just clay pipes still with evidence of beeswax inside. Over these years beehives have been invented and improved but bees themselves have not changed. The hive, the house where the beekeeper keeps his bees, is for his convenience, not for the good of the bees. Bees are happy in a hollow log where they have always lived, but the beekeeper needs to get inside the hive, to see the bees at work, to check them for disease and eventually take some of their honey.

This is a very old straw hive, used long ago. It does not allow the beekeeper to inspect the bees and harvesting honey involves damaging the colony to remove the combs.

Hives like these are more familiar now and have been in use for over 150 years.

Inside the boxes are wooden frames in which the bees build their combs.

The beekeeper can remove frames from the hive and see the bees and their honeycombs.

Harvesting honey in times long gone involved major damage to the bee colony, even complete destruction, treating bees and honey like annual crops. Getting honey from early hives, including the straw skep hive, necessitated removing honeycombs and major disruption of the bee colony. It fell to a clergyman, Lorenzo Langstroth, a Congregational parson in Philadelphia, USA, in the first half of the 19th century, to design a hive which gave the beekeeper access with minimal damage to the bees and their combs.

Langstroth based his design on his studies of what he called "bee space"; the 9-10 mm space bees require to move around and work their combs. Give them more than that and they will build more comb in it, inaccessible to the beekeeper. Langstroth's hive provided removable hanging wooden frames just that far apart so that combs are built neatly within them and not all over the place . This enables bees to move freely and makes it possible for the beekeeper to remove individual frames of honeycomb to inspect the health of the colony and ultimately to harvest honey. To ensure that the bees build nice straight combs in and not between the frames, the beekeeper puts thin sheets of beeswax imprinted with the hexagonal pattern of honeycomb, called foundation comb, in the frames. The bees accept this template and build their comb on it. Newly made comb is off white in colour, becoming darker as it ages, eventually turning nearly black.

A recent innovative Australian development in beehive design, called the Flow Hive, has made harvesting honey very simple, by turning on a tap at the back of the hives. Stuart and Cedar Anderson, father and son, developed a hive box with plastic frames and combs which can be opened inside by an outside lever mechanism. The comb is fractured internally, the honey drips to the bottom of the box whence it can be drained to the exterior and collected from a tap directly into a jar without disturbing the bees. It must be emphasised, however, that this innovation only involves the honey box, the upper deck of the hive or "super". The importance of the brood box below is unchanged and the care of the bees is just as demanding. So far its use is appropriate only for amateurs or beekeepers with a small number of hives, but the invention has given amateur beekeeping a world-wide boost.

A frame with foundation comb

There are many variations on the Langstroth hive, Top Bar hives, Warrè hives, but if you keep honey bees it is a legal requirement to use removable frames so combs can be inspected for disease.

A homemade Warrè Hive

Honeycomb is used by the bees for three purposes: to rear brood, to store pollen and to store honey. In an unmanaged hive, the natural state in the wild, all three of these may be found in one comb, which is fine for the bees, but not for the beekeepers who want to harvest honey without disturbing the all-important brood or getting pollen mixed with honey. As a managed colony gains strength it needs additional space, the beekeeper obliges by adding new boxes with frames loaded with foundation comb, one on top of the other, sometimes four storeys high. These are called "supers". To ensure that these supers will not be used by the queen for egg laying, beekeepers play a mean trick. Above the bottom box they put a queen excluder, a mesh grid with holes too narrow for the larger queen (and even larger drones) to get through but allowing access to the smaller worker bees. So these supers contain no brood, just honey stored there by the workers and accessible to the beekeeper.

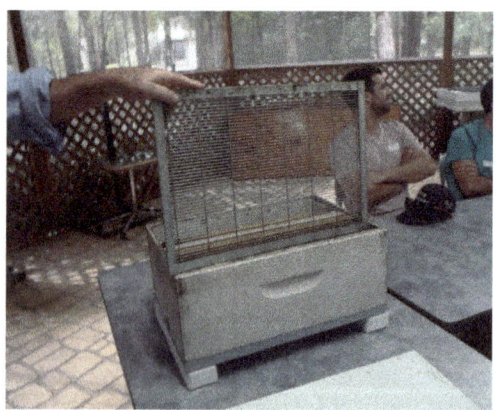

A backyard beekeeper's three deck hive. The black strip above the bottom box is the queen excluder.

With the queen confined to the bottom box, this is a nursery, called the "brood box". The bees do store some honey there, but that is for their use only. In the upper box(es) there is only honey. So we could say that the brood box is for the bees and the supers for the beekeeper, but that is not quite true. Remember that honey is the carbohydrate food for the bees and they need a lot of it to supply energy for all the work they do. The imponderable miracle of nature is why this honey bee makes far more than immediately needed.

Other species of bees don't. So the wise beekeeper never takes all the honey in a hive, not even all the honey above the queen excluder. He knows that in warm climates he must leave at least half a box full of honey for winter feed, and in colder areas a full box.

Harvesting Honey

When will the beekeeper harvest honey? When there is a good honey flow, plenty of honey coming in, and the honey super or supers are near full. This will be in spring and summer, depending on the available floral sources. With good strong colonies and good honey flow from abundant flowering trees and plants harvesting may be done twice or more in a season, with strong colonies in good times yielding 20 or more kilograms of honey each time.

Primitive methods (as with the skep hive), involved removing full honeycombs and squeezing the honey out from the wax, so destroying the combs. Modern methods allow extraction of the honey from the combs with minimal damage so they can be put back in the hive for the bees to refill[7]. Rebuilding honeycomb requires a lot of energy to be expended by the bees so saving the combs is a great advantage to bees and beekeeper. Eventually, of course, the combs do become damaged or old and must be removed. The beekeeper will replace the frame with a new one with a wax foundation sheet for the bees to build completely new honeycomb.

---

7 I love to tell children that the beekeeper puts the combs back with a note to the bees saying, "Please refill". And bees must be able to red because they do!

Getting the honey from the hive

We used to talk about "robbing bees" and that's what we really do because they store honey for their own food, but now the term "harvesting" is preferred. Fortunately for us bees store more honey than they need so we can help ourselves to some of it. But how does the beekeeper get the honey out of the hive?

These pictures tell the story.

The hive is opened

Honeycombs are removed

The wax lids are cut off the combs

The combs on the wooden frames are put into a spinner, called an "extractor"

And the honey is spun out

Then the beekeeper puts the honeycombs back into the hive, now with empty cells drained of honey and ready for the bees to refill.

Professional commercial beekeepers use modern technology and sophisticated equipment for their operations. With so many hives, mechanical plants are required to uncap the combs and extract the honey from hundreds of frames.

Amateur and Professional Beekeepers

An "amateur" is someone who does something as a hobby, for fun or recreation, like a sport. A professional person is one who does it to earn money, to make a living. And so there are amateur beekeepers (sometimes called "recreational beekeepers") and professional beekeepers, (also called "commercial beekeepers").

The amateur, or hobbyist, has one, two or more hives, but not many. He puts his bees in one place, usually his backyard, and tells the bees to "go find your own flowers". And they do. Bees can be kept in towns and cities and hundreds of hives are found there. There are plenty of flowers in gardens and parks, and bees don't have trouble finding them. Amateurs have a "code of practice", a set of rules to guide them so their bees do not upset neighbours. And neighbours just *love* the jars of honey they are given.

Author's backyard in summer with shade cloth protecting hives

A commercial apiary owned by Anthony Pyne who supplied the photograph

A professional beekeeper with truck loads of bees going to where there is a flow of nectar

Photograph from Roberts family of Rainforest Honey

To make a living from bees the professional must to have over 200 hives, some have a couple of thousand. He takes his bees on a truck to where trees or crops are in flower, and when the flowering is over he picks them all up and takes them to other floral sources elsewhere. So professional beekeepers must know as much about flowers, trees and crops as they do about bees. They use modern technology and sophisticated equipment for their operations. With so many hives, mechanical plants are required to uncap the combs and extract the honey from hundreds of frames.

Commercial extraction and processing honey is highly mechanised.

Photographs by Bill Winner

Management of bees by beekeepers, Caring for bees

There is far more to beekeeping than collecting honey. It is true that bees have been looking after themselves in the wild for thousands of years, but exploitation for honey production requires care and attention by beekeepers who need to learn good management practices. If they are to be successful they must look to the health and welfare of their bees by identifying and dealing with diseases and pests, ensuring adequate nutrition, adequate space for breeding and storage, appropriate temperature control and productivity.

Beekeeping in Australia is strictly controlled by the State Government Departments of Primary Industries with which all keepers of the honey bee, amateur and professional, must register. A number is allocated and this must be displayed on all hives. (Australian native bees do not have to be registered.) A Biosecurity Code details the level of care required, demanding a complete inspection of the colony at least twice a year. This is a minimum; it is good practice to examine the whole hive monthly from spring to autumn, with particular attention to the brood box. Methods of detection and control of disease are specified, and in the case of American foul brood particularly, they are legally enforced.

For maximum productivity the strength of a colony must be maintained. Controlling disease and pests and ensuring nutrition are obvious essentials, but attention to the biology of bees is also important. One vital key to the strength and productivity of a colony is the health of the queen. It has already been noted that a queen bee in the natural environment lives several years, three, four or even five, but like the females of most species, she loses her fertility as she ages. Left to themselves, bees will detect her failing by the fall in pheromone level in the hive and they will breed a new queen to supersede the old one. Provided they have one of her eggs of newly hatched larvae, they will build a large queen cell around it, feed the larva with extra royal jelly and cap the developing pupa to await the emergence of the young virgin queen in 16 days. They will usually make several new queens, but the first one to emerge will destroy the others by stinging through the capped cell. If two emerge together they will fight to the death. This new queen is unmated. She rests a while, then flies out on mating flights to meet with drones from far away, from other colonies. She will mate with 15-30 drones, storing the accumulated sperm for the rest of her life. (The process of mating and the significance of multiple mating has been described in a previous chapter, Reproductive Biology.)

From the viewpoint of a beekeeper, this natural process of queen replacement has disadvantages. There is delay as the bees become aware of the failing queen and more time is lost till her new progeny emerge to work. Further, the queen mates with drones of unknown quality and these drones, of course, supply half the genetic material which determines the quality of the worker offspring. So beekeepers take the matter into their own hands and replace a queen *before* she fails. Commercial beekeepers whose very livelihood depends on high productivity replace the queens every year; amateurs, not under so much pressure, do it when they get around to it, usually every two years.

So how do beekeepers replace a queen?

First, queens must be produced. Some beekeepers breed their own queens but most depend on professional queen breeders who sell their queens for varying prices, at the time of writing about $20-30.

How do you breed a queen?

You trick the bees into doing it for you by making a colony queenless. But breeders are mean and trick the bees into making many queens at once. By a technique called "grafting" they take newly hatched larvae from a well selected parent hive with a known high quality (mother) queen and put them individually into multiple cells on frames. The frames go into a queenless hive and the bees obligingly do the rest. But, knowing that the first queen out will kill the rest, the breeder removes the cells just before emergence of the queens and puts them individually into other queenless hives called nucleus colonies or "nucs". There the queen emerges safely, goes through the mating process, returns and starts laying eggs. It is then that the breeder catches and cages her for sale. And what happens to the "nuc" now queenless? It can be used for another round of grafted queen cells, allowed to requeen itself or amalgamated with a hive with a queen.

A bar of queen cells. The queen breeder provides the plastic cups, the bees make the wax cells. They always hang vertically downwards.

Photograph supplied by DPI

There is one other demand on the queen breeder. We have noted that mating with "quality drones" is important and breeders must pay equal attention to the availability of such drones in their area of operation. Drones congregate in particular areas, flying around on their strong wings awaiting the arrival of a virgin queen which they can readily spot with their big eyes. We have already noted that the successful fellows die after rendering their service.

How would you find a drone congregation area? You take a caged virgin queen, attach to a balloon on a long string and search likely areas in the vicinity of an apiary known for the quality drones. Once found, the drones will zoom on to the cage in numbers. The poor fellows are frustrated but their lives are saved!

Photograph by Tom Gillard

Then, having bought his newly bred queen how does a beekeeper introduce her to his hive? Remembering that there can be only one queen in a colony, the beekeeper must first kill the old queen, and to do that he must first find her; one bee, only a little larger than the many thousands of worker bees. He knows that in a managed hive with queen excluder above the brood box she will be somewhere there in that bottom deck, but still just one bee amongst many others. So he starts on one side of the box and, removing and inspecting

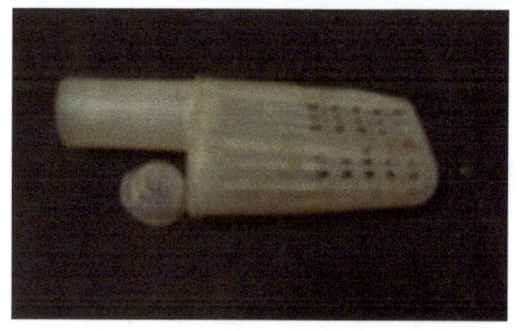

frame by frame and till he gets to the other side. If he hasn't found her, he starts again! Four eyes are better than two for this task. Eventually the cry goes up, "There she is!" And the poor girl is grabbed and squished, one of the hardest things for a novice to do for it seems such a pity to destroy such a beautiful creature.

The newly bought queen is in a cage with an exit spout plugged with candy made from sugar and honey. She is accompanied by several worker bees, called escorts, to feed her from the candy. The cage is inserted into the brood box frames and in the next few days the hive bees eat away the candy from without the cage and the escort bees eat it from within. Meantime the hive bees have realised that their old queen is dead and learn to accept the smell and pheromones of the introduced queen. During this critical time the beekeeper leaves the hive well alone, refraining from inspecting it for at least a week. When he does he hopes to see eggs and young larvae in the combs, confirming that the queen has been released and is laying.

Requeening like this has a 95% success rate. What happens to the other 5% where the queen was rejected, damaged or killed in the process? The beekeeper can try again, or if another queen is not available he can put a comb with eggs and young larvae from another hive so the bees can make their own queen, but with the time this takes the colony will weaken significantly.

Beekeepers have a trick to make it easier to find a queen. Once found she is marked on the back with a coloured pen. Using different coloured markings makes it possible to tell at a glance how old the queen is.

Photograph by Des Cannon

The code is:

White mark for queens introduced in years ending in 1, eg 2001, or 6, eg 2016

Yellow mark for queens introduced in years ending in 2 or 7

Red mark for queens introduced in years ending in 3 or 8

Green mark for queens introduced in years ending in 4 or 9

Blue mark for queens introduced in years ending in 5 or 0

(Remember that no queen will live more than 5 years, so there is no need for separate colours for years beyond five. You do not have to remember this but beekeepers do. Some have difficulty, so they use a mnemonic such as: *When You Retire Get Bees.*)

A cute story from talking to small children about bees and showing a queen marked with a white spot on her back: "How do you think that white dot got there?" Asked the beekeeper. A sweet faced little kindy looked wistfully into his eyes and suggested, "A bird dropped a poo on her?"

Photograph by Des Cannon

This very dark queen with her equally dark worker bees is Carniolan originating from Western Europe. She is not only marked with a colour but with a number. You will notice that she (like the one pictured above) has 0ne wing clipped. This is a practice of some beekeepers to prevent her flying and hopefully to stop the colony swarming.

The queen bee is the only live animal which can be sent in the post as illustrated by this old queen mailing cage.

Picture from "Bees and Honey"
WA Goodacre, "Bees and Honey",
NSW Department of Agriculture, 1941

Successful requeening of a colony is a most fascinating and rewarding experience. The beekeeper can actually *see* and *feel* the benefits. From day to day as one studies the foraging bees returning to the landing board of the hive, changes in colour may be evident. If the old queen was dark (Caucasian or Carniolan race) and the new one a golden Italian (Ligurian race), after three weeks an increasing number of more golden bees will be seen as her progeny replace the old queen's offspring, and eventually they will all be of the new colour. The "feel" of the success of the requeening exercise will be in the changing temperament of the bees. Assuming the new queen was bred from quiet non-aggressive stock, (and hopefully they always are) the beekeeper enjoys the ability to manipulate the hive without being attacked by myriads of suicide bombers!

At a beekeepers' field day a queen breeder was demonstrating his skills in the open paddock. People watching, including myself, noted with envy that he needed no protection, not even a veil; the bees staying calmly on the combs making no attempt to attack. All the observers said, "*I want some of those queens!*" I did buy some and set about to requeen three hives I looked after for a friend on a country farm. The previous beekeeper had died and I had inherited very aggressive bees. I had to wear maximum protection when going near them and cattle venturing close were sent off with a bellow and a swishing tail! After requeening with quiet stock I could open the hive with no more than a veil. I'll leave you to imagine the challenge and "fun" involved in finding the old queen in such aggressive colonies!

There is a lot to learn to be a successful beekeeper, whether amateur or professional. Books provide information, but there is more to beekeeping than book knowledge. One must gain confidence to handle bees, to care for them, to detect and deal with pests and diseases, and this can be achieved only by hands-on experience guided by more experienced beekeepers. Laws, regulations and biosecurity issues that must be familiarised. Amateurs are well advised to join their local branch of an Amateur Beekeepers' Association. In NSW, Australia, there are many branches stretching from Bega in the south to Northern Rivers.[8]

---

8 See ABA website: www.beekeepers.asn.au

HONEY

Ned says "yum"!

Photograph by Kate Connor, Ned's Mum

Honey is good food. People have been eating it for thousands of years. It is a sweet sugar made from nectar, not pollen. Honey lasts forever. Honey found in Egyptian tombs from thousands of years ago was said to be still edible.

Stored honey exists in three forms: liquid, candied (crystallised), and creamed. Liquid honey varies in density as well as colour, and to test the density invert the bottle and see how quickly the air bubble rises. It will candy (crystallise) if refrigerated but can be frozen even with comb in it. Candied honey is due to crystallisation of the sugars, is temperature dependent and occurs quickly at 14°C, depending on the proportions of sugar in it. It can be coarse grained or smooth grained depending on the floral origin. Honey from some floral sources candy more quickly than others. Creamed honey is made by aerating a mixture of 10% candied honey and 90% of liquid honey using a strong blender-mixing machine. Never use an ordinary kitchen blender because honey is too thick. ("Creamed honey" is really a misnomer, it should be called "whipped honey".) This form of honey must be kept in the refrigerator to prevent it reverting to liquid.

The taste of honey depends on the flowers the nectar came from, not on the bees collecting it.

Other good things are claimed for honey as a food. It is said to be good brain food; if you eat it before bed it goes on feeding your brain through the night. And it is claimed that it does not rot your teeth. It has also been proven to be a good cough medicine, better that a bottle of cough mixture from the chemist.

Honey is made from the sucrose of nectar by two main processes; break down of the carbohydrate sucrose to glucose and fructose by enzymes provided by the bee, and removal of water.

Nectar is 70-80% water, honey 17%-19%. Bees remove the water by passing the nectar from bee to bee and by drying it out and depositing it in the cells. The bees can tell when it is sufficiently dried out and thus when to cap the cell. Immature honey with too much water will ferment, mature honey will not, but it is hygroscopic[9] and will attract water again if left open.

Mature honey consists of:

    Water: 17%

    Sugars: Fructose (fruit sugar) 44%, Glucose 35%, Sucrose 3%,

    Other solids including residual pollen: 1%

The colour of honey depends on the floral species sourced but it will darken if heated above 43°C. Such heat will also modify the content compromising its nutritional value and destroy its antibacterial properties. (See below.)

As a food, honey is an excellent source of carbohydrate with an energy value of 1230kJ per 100 grams. It is also an effective "prebiotic"[10] feeding and encouraging the growth of "good' bacteria in the gut.

---

9   Hygroscopic: drawing water from surroundings.
10   A "prebiotic" is a food substance which encourages the growth of "good" bacteria the gut and this helps control the growth of harmful (pathogenic) bacteria.
     A "probiotic" is a culture of living non-pathogenic organisms in the diet; for example the live bacteria in yoghurt.

Fermented Honey

Mature honey will not ferment. It does not really have an expiry date; it can be kept indefinitely. But if it is diluted with water, it will ferment and produce alcohol from its sugars. From time immemorial this has been exploited to make the alcoholic drink, Honey Mead. It is a drink to be sipped before the fire on cold winter nights!

A family living in a wooded area loved wild birds which they fed generously including putting honey in their water bowls. When told by their beekeeper honey supplier that honey so diluted would ferment to alcohol, the response was, "So that's why they fly in quickly in perfect control and fly out slowly and all wobbly!"

Honey as an antiseptic

Used as an external application honey has antiseptic properties, it kills bacteria. It is not an antibiotic and does not cure illnesses when taken orally. Its use over centuries as a dressing for wounds and ulcers was forgotten when antibiotics were discovered and were expected to cure everything. Time has proved that this hope was ill founded and the antiseptic properties of honey have been rediscovered and are again being exploited. All honey that has not been heated in processing has some antiseptic qualities but its strength varies depending on its floral source. Extensive research is directed to identifying these sources and the content which provides the activity.

Flowers of the Leptospermum[11] variety produce the most active antiseptic honey. The New Zealand beekeeping industry was first and most active in this field and their Manuka honey is so well known that the name is now applied to medicinal honey generally. "Manuka" is one of New Zealand's three species of Leptospermum. Australia has about 83 species but the honey they produce varies widely in antibacterial potency.

---

11 Leptospermum means "tiny seeds". The plant is otherwise known as "jelly bush" and is of the Tea Tree variety.

Flower of a Leptospermum species

Picture from Wikipedia

Leptospermum polygalifolium[12]

The antibacterial activity of honey is due to several factors:

▸ It is highly concentrated, hypertonic and hygroscopic which makes it attract water, drying bacteria, making them like shriveled prunes, or as one writer expressed it, transforming them "from grapes to raisins"

▸ It is significantly acid, pH 3.4--6.1

---

12 Polygalifolium: poly *many*, folium *leaves*, gali *has to do with milk*, because the leaves resemble the plant milkwort which was fed to cattle in the belief that it increased milk production.

- Enzymes in it produce hydrogen peroxide which is antibacterial
- Floral specific content identified as methylglyoxal ("MGO"). The concentration of MGO is used as a measure of the honey's level of efficacy.

As a dressing on ulcers, open and infected wounds honey has other advantages:
- The dressings don't stick to the wound making removal less painful
- It removes the smell of infection
- It has a desloughing effect, removing dead tissue.

Leptospermum honey is much thicker, more viscous than honey from other floral sources, making extraction from honeycombs more difficult. There is no health advantage in eating medihoney, which is just as well because it is understandably more expensive than ordinary honey.

POLLINATION

Pollination is the process of transferring pollen from the male part of the flower, (stamen), to the female parts (stigma and style) of the same or other flowers of the same species.

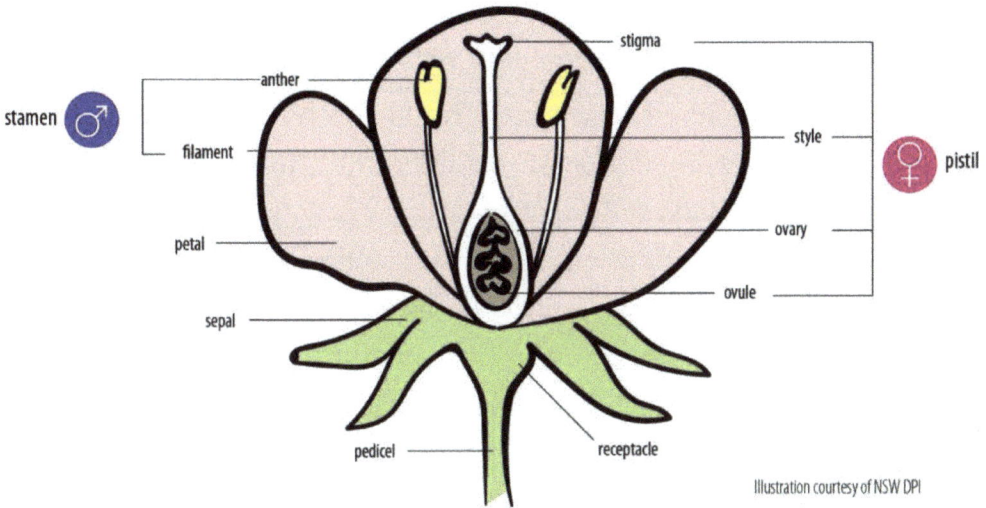

This is a diagram of what is inside a flower. The bees transfer pollen from the male part of the flower (left side) to the female part (right side)

Department of Primary Industries, courtesy of Elizabeth Frost

Pollen is the male gamete of flowering plants, plant sperm. This fertilisation process is essential for fruit to develop and seeds to form. One in every three bites of food we take depends on pollination. Nature has several methods of doing this: wind, water, gravity, animals and insects. Yes, animals do contribute to pollination. It has been discovered recently that bush rats pollinate Banksia. Even humans contribute with hand pollination in times of need, but their efforts are very inefficient and time consuming, demanding a large poorly paid workforce and infinite patience. Mankind cannot compete with insects which contribute 90% of animal pollination. Of these, bees have a prime role with 75% of food crops depending upon them. Of course, Australian native bees pollinated Australian flowering plants before the arrival of the European honey bee in 1822, but there were no massive food crops requiring their service. It has been said (but only *said*, not proven) that if we lost the honey bee, mankind would not survive more than a few years.

Pollination occurs between flowers of the same species. Plants and animals of the same species can interbreed, though there are different races within species. (Think of Homo sapiens, all one species but with many races which can interbreed.) It is remarkable, and a wonder of nature, that on each foraging flight a bee visits flowers of the same species.

But bees must have access to a variety of flowers because pollen provides nutritional protein, fat and vitamin requirements. No one pollen supplies all their nutritional needs, so bees must visit a variety of flowers on different flights, and the beekeeper must see different colours in the pollen baskets on the hind legs of foraging bees returning to the hive. Pollen consists of tiny granules which are distinctive in shape and microscopic appearance enabling scientists called palynologists [13] to identify the floral source of honey from the few residual pollen grains in it. *(If there are no pollen grains in honey it is either artificial honey or it has been ultra-filtered.)*

Professional beekeepers with their hundreds of hives are in demand by farmers who need bees to pollinate their enormous crops and orchards. They pay beekeepers to put their hives near their farms and orchards when the plants and trees are in bloom. Commercial beekeepers nowadays need income from this source as well as from the sale of honey.

---

13 The study of pollen residue in honey is called "melissopalynology"

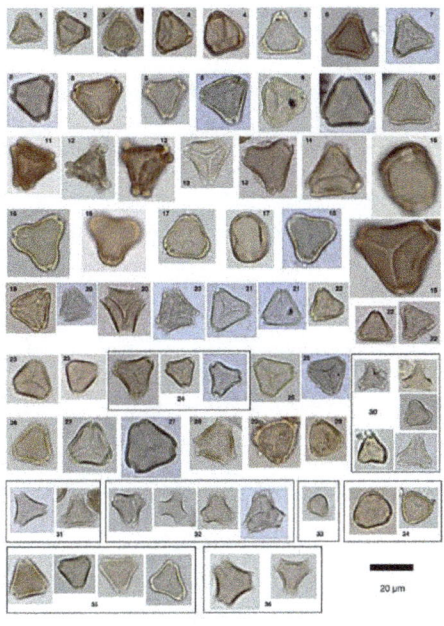

Pollen Grains

Picture courtesy of Professor Simon Haberle, Australian National University

MEN OF THE CLOTH (Clergy)
involved in Beekeeping

This is a subject of general interest. It is remarkable that men of the cloth, monks, priests and clergy have played a prominent role in the history of beekeeping. Long before refined sugar, honey was the universal sweetener. And the monks loved their alcoholic mead made from honey. Can you picture a group of them before the fire on a freezing winter night, sipping warm wholesome honey mead? Beeswax was also in demand by clergy to make candles because beeswax candles do not emit smoke to stain the roof of the church.

Those early clerical beekeepers no doubt used primitive hives like the straw skep hive. It was a Congregational parson in Philadelphia, Lorenzo Langstroth, who invented the modern hive and patented it in 1852.

Picture from Wikipedia free images

Later in the same century, another cleric in France, Abbe Emile Warré, simplified the hive to make it cheaper for poorer beekeepers. In fact, he called it "Ruche Populaire" (people's beehive). This Warré hive is still used today by some amateurs who espouse more "natural" beekeeping practices, but it also necessitates sacrificing the combs when honey is harvested; not appropriate for commercial beekeeping.

Picture from Wikipedia free images

On to the 20th century and we find another monk in England contributing so much to beekeeping in that country that he was awarded an OBE[14] in 1974. Brother Adam, a Benedictine monk of the Buckfast Abbey in south-west England, was responsible for breeding bees, known as the "Buckfast bee", resistant to a parasitic Tracheal mite which had wiped out local indigenous

---

14 OBE: Order of the British Empire, equivalent to an Order of Australia award.

bees. He was an intrepid soul. I saw a video of him working with vicious African bees with no protection. Without any sign of concern he turned to the camera and remarked, "They are stinging me!"

Brother Adam

Pictures from Wikipedia free images

Perhaps we could add the Australian "flogging parson", Samuel Marsden, to this list given his early attempt to bring the European bee to Australia.

 Rev Samuel Marsden

It is salutary to note that these clerical beekeepers lived to ripe old ages; Langstroth 85, Warré 84 and Brother Adam 98. We should take note of this and follow the advice of a more ancient ecclesiastical writer: *"My son, eat thou honey, because it is good; and the honeycomb, which sweet to thy taste."*.[15] Perhaps Rev Samuel Marsden did not heed this advice for he died at the age of 73.

---

15   Proverbs 4 : 13

Who is the Patron Saint of Beekeeping?

It is **Saint Ambrose,** but few people would have heard of him. He lived in Italy during the late Roman Empire after Christianity had become the established religion. A successful governor of one of the provinces in Italy, he was asked to become bishop when the incumbent died and after some persuasion he agreed.

At that time the Church was embroiled in a theological dispute between a conservative view and another held by people called Arians who were later branded heretics. Ambrose took the conservative view and helped push the Arians off the stage. Ambrose also had some views about the separate roles of Church and State in government and was quite free with his advice to the Emperor at the time. His activity in Church affairs and government was probably the main reason for being made a saint.

His connection with beekeeping was very slim. It was said that as a baby asleep in his father's courtyard bees gathered around him and several were seen to go in and out of his open mouth without waking him. This was regarded as an omen to the effect that he would say many sweet things in later life. History does not record whether he did or not.

HOW WAS THE EUROPEAN HONEY BEE BROUGHT TO AUSTRALIA?

The European honeybee was an early immigrant to Australia. No doubt settlers soon discovered that Australian native bees were not sufficiently productive to allow harvesting of honey and endeavoured to bring out the bee familiar to them, the European honey bee, *Apis mellifera*. After two failed attempts by famous early settlers, Samuel Marsden in 1805-1806, and Gregory Blaxland (later of crossing the Blue Mountains fame) in 1810, the honey bee was successfully lodged in Hyde Park area, Sydney, in 1822, after a long voyage from Ireland in the ship *Isabella* under the command of Captain Wallis. Details of how this extraordinary feat was accomplished are obscure. Given that the life of a bee is about six weeks, though more in the inactive state aboard ship, there must have been breeding activity during the trip. Feeding the bees would not have been a problem but confining them on board ship would have been. Being hygienic creatures, bees do not defaecate within their hive and must fly out to do it. So perhaps they were given netted space on the ship. Early voyages were staged, (Canary Islands, Cape Town), but it is recorded that the *Isabella* came direct from its first and only stop-over in Ireland, leaving there on 4th November 1821 and reaching Sydney Cove on 19th March 1822, a seventeen and a half week voyage, nearly three times the ordinary life of a bee.

There are conflicting reports of the number of hives loaded on board, 4-10 being quoted. At least four survived the journey and landed in Sydney. The last of these, purchased by Mr Parr, of George Street, Sydney, is alleged to contain 476 bees! How did he know? Did he count them?

Let your imagination reflect on the "thoughts" of the bees as they were released to discover the rich nectar of the Australian Eucalypts; they must have been ecstatic! They certainly thrived and were gradually moved west, though records indicate that it was 1839 before they were established in Bathurst, 26 years after the initial crossing of the Blue Mountains. Elizabeth Macarthur played a part in this westward movement.

The variety of *Apis mellifera* originally introduced was the black English/German race, *Apis mellifera mellifera*. The golden Italian bee, *Apis mellifera ligustica* was subsequently brought out in 1862.

Considering the logistic and biological difficulties involved in this exercise, we are left with deep admiration for the ingenuity and determination of our forebears, overcoming as they did the innumerable problems involved in the transport of bees over thousands of miles with several months at sea. Because of their achievement we have an apiculture industry, pollinators for our major crops, a hobby to pursue and honey to enjoy.

ARE BEES DISAPPEARING?

Serious Threats to Bees

You may well be aware of concern expressed in the news media about dangers to bees in the world today. This is very real, and very important. We have already looked at how bees pollinate flowers and crops that produce our food. Bees are not the only pollinators; other insects, birds, wind and even gravity play a part, but more than half our food crops depend on bees. So if we lose bees, our food supply is endangered.

So are bee numbers decreasing? Most people agree that they are, and there are many causes. We noted the disastrous effect of diseases and parasites which are of special significance in the northern hemisphere, America and Europe. Production of our food, fruit, cereals and vegetables

now involves huge crops many hectares in size. In contrast to bees, many wild insects are a menace to these crops and farmers use insecticide poisons to control them. But bees are insects too, and they suffer.

In the wild, out there in nature, bees live in forests, but we, mankind, are cutting down forests at an alarming rate. And if you look around you in a town or city, do you see many flowers? Concrete covers so much and where there is room for a flower garden you will see ground covered with wood chips featuring non-flowering plants. We don't like weeds, but bees love flowering weeds[16].

Bees are being denied food in the bush, cities, towns and home gardens.

Around my suburban villa the grounds have been sprayed for spiders. There are no spiders but there are no native bees either. Put there in the hope of providing a place for native bees to pupate, this native bee "hotel" remains conspicuously empty.

---

16 What is a weed? It is just a plant in the wrong place, wrong for us but not necessarily in nature.

We trim our hedges to look neat like this, but where are the flowers?

A recently replanted garden area in a suburban mall. These plants do not flower. Note the woodchip covering the ground so no flowering weeds can grow and ground breeding solitary bees cannot pupate.

This is typical of development in town and city areas. From bush land with habitat and food for bees to the wasteland of a construction site, soon to be a covered with sterile concrete.

Is it any wonder that people living in these environments ask, "Where have the bees gone?" So if you love bees, love flowers; they depend on each other.

MEMORIES OF A DYING BEE

This story tells in fantasy the sequence of duties of a worker bee in her six weeks of life.

A worn-out old honeybee lay dying

No pillow supporting her head

No other bees around her a'crying

If she could speak she'd have said:

"I remember.........................."

If it is true that in dying moments a kaleidoscopic view one's life flashes through consciousness, this is what a dying honeybee might experience. Bear in mind that a bee lives only about six weeks and literally works herself to death, flying until wings become too tattered to carry her home. And if you've seen such a bee on the ground writhing and squirming under attack by ants you will feel a pang of sympathy for this little creature and perhaps shed a tear for her as she thinks:

"I remember so long ago, six whole eventful weeks, when I chewed my way out of my cell into the busy activity of my family, my fifty thousand siblings. I had to start work right away. Cleaning was my first job and where better to start than with my own birth place, the cell from which I emerged. Of course, I didn't have to go to school; we don't need teachers because nature has programmed us to act instinctively. I knew I was not alone because more than a thousand others were born the same day. There was plenty more work to do; my older sisters making a bit of mess for us youngsters to clean up after them.

I soon outgrew this menial task and moved to my next job. I became a dietary maid, a nutritionist. I had to feed the babies. To do that I had to mix honey and pollen and make lovely rich royal jelly, giving just a little to each tiny white baby to make it grow. As a senior dietary worker I had the greatest privilege; I had to feed our lovely queen, my mother. This was wonderful, getting so close to her and sucking up her pheromones which I then spread around the rest of the colony. I also had the pleasure of grooming her which brought me even more under the powerful influence of the pheromones which she exuded. I thought my queen was beautiful and admired her for her ability to lay egg after egg, more than a thousand a day!

I also had to feed the drones; those big good-for-nothing males with eyes that meet in the middle, broad shoulders and big hairy backsides. They did nothing in the hive, but would go outside periodically, fly around and come back. I noticed that occasionally one did not come back and word went around in whispers that he had met with a virgin queen and mated with her, unfortunately dying immediately afterwards. Tough!

A less exciting but nevertheless important role followed. I had to receive sweet nectar from my older sisters who had been outside the hive collecting it from flowers. As they transferred it to me I helped dry out the excess water and added my enzymes to it to break down the sucrose into glucose and the lovely sweet fructose. Then I put it neatly into a cell and fanned it to dry out more water. I was actually making honey!

My next job was also interesting and responsible. I learned to make wax from the glands under the scales on my abdomen. I learned to scrape it to my mouth, to chew it, mix it with my other body juices and then make wonderful things with it. First it was my job to make the caps on the cells containing honey and the developing larvae, soon to be my younger sisters. Then I became really important; I felt like a construction engineer making whole honey combs, hundreds of lovely new wax cells, all perfect hexagons. Of course, this was a team effort and it was great working with so many others, holding on to them to form a sort of scaffold to build on.

So far all my duties had kept me house bound—except, of course, for cleansing flights. Unlike some other insects I've heard about, we bees are a clean creatures; we never sully our own nest and always go outside on a short flight to cleanse ourselves. I remember one time when a car was parked in front of our home. My sisters and I left our faeces all over it, like yellow rain it was, sticky yellow dots very hard to remove.

Getting older now, I moved to my next duty. I had the task of ventilating the hive to keep it cool and dry out the nectar in the process of making honey. It was fun! I'd stand still in the entrance to our hive and fan my wings like crazy; on one side blowing air in or on the other drawing it out. It was a trick to flap my four wings so fast without flying, like staying in neutral gear!

Then getting still older I was assigned guard duties. Patrolling the entrance, I had to watch out for invaders, bees from foreign colonies who might have sneaked in to steal our precious honey, beetles or those other insects like us but awfully nasty, wasps. There was also a human who burst in to steal our honey, but he cheated; he blew smoke over us and so confused us so that we could not put up a fight. One of my friends did put up a fight and stung him, but she could not get her sting out. She pulled and pulled till her bottom came off and some of her innards came with it. It was awful to see her fall away to die. Not a pretty sight, and all to no avail because the man still got our honey! I'm so glad I wasn't called upon to make that ultimate sacrifice and was able to live out my full life span, though I'm finished now.

It seemed that all these early days were in preparation for my life's last task. I was old enough to leave the hive to forage for food in the outside world. Of course I had to do some orientation flights, to learn landmarks and how to use them and the position of the sun to navigate. It was good to know that I could still rely on the sun despite cloud cover because I was sensitive to ultra-violet light. My experienced older sisters already outside scouting and foraging gave me a good head start. When they found a good mass of flowers they came back and told us all about it by fancy dances. We quickly knew the direction and distance from home and how rich the supply of nectar or pollen. They even left their pheromone smell at the site to act as a guiding beacon. Sometimes I went out for sweet nectar, sucking it up from the flowers with my long tongue, storing it in my honey stomach till I could pass it on to home bees back in the hive. Other times I looked for pollen, scraping it into the little sacs on my hind legs. Or water; we needed water in the hive and getting it was sometimes hazardous. Many of my sisters drowned in swimming pools trying to get water from the surface. Better to go for bird baths and sit on the edge sucking it up from the wet concrete. Even the edge of muddy pools was better and much safer.

I flew many a mile on these trips, up to three or more kilometres when I had to. On each trip I'd visit over a hundred flowers. I could carry only 30-50 milligrams of nectar each trip and it got pretty heavy after a long flight. I was often so tired I'd sometimes fall short of the landing strip back home! And to think that it took 400 of those trips to make one teaspoon of honey! No wonder I had so many sisters in the colony to share the work load.

I wonder how many times I beat my little wings in my life time? I could beat them 200 times a second, 1,2000 times a minute! No wonder they are all worn out now!

And that's why I'm here, lying in the dirt far from home. My wings finally tore as I tried to work on that last flower and they just will not carry me home. So here I wait to die very soon. No bee of my species can live alone outside the hive, we are just part of a big corporate body and all that matters is the good of the colony. I wonder if I'll go to a bee heaven? If I do I'll meet many millions of my kindred there. But right now I just hope that I die before the ants find me in my helpless state. They are a cruel lot and will chew and bite me mercilessly. But that is the inevitable fate after all those hard working days devoted entirely to the service of my huge family.

Good-bye cruel world!"

Vale little bee. Requiem.

AUSTRALIAN NATIVE BEES

The honey bee we have spent so much time on is not a native Australian, it is of European origin and was brought here by ship in 1822. But Australia has its own bees, over 1,500 species, or types of them. If you look very carefully you will see some of them in the garden. They are small and move very fast. Some, like the European honey bee *Apis mellifera*, live in colonies. Like the bees themselves these colonies are much smaller than those of the honey bee and the hives used to keep them are little boxes with a hole in the front. They do not fly very far, half a kilometre at the most. And they are very sensitive to cold and extreme hot weather.

Artistry and Poster by Gina Cranson

Best known is the Australian Stingless bee, *Tetragonula carbonaria*; tiny, black and with a white face (top right hand corner in the poster above). These are good pollinators and many people are keeping them in their backyards to pollinate and improve production from their gardens. They do store honey but nowhere near as much as the honeybee. It can be harvested but one would be lucky to get a kilogram a year. It is darker and "stronger" in taste than honey bee honey; very nice with icecream!

Painting of the Australian stingless bee, *Tetragonula carbonaria*

by Gina Cranson

This is a native bee hive

Photograph by the author

This is what a native beehive looks like inside

Photograph by the author

The nest is made up of comb built in a spiral rather than sheets like the honeybee.

Photograph by Anthony Pyne

This shows beekeepers working with an open hive of native bees. You will notice that one has a veil. Though these little bees don't sting, they crawl into your eyes, ears and hair. The man with a veil has a beard and bees love to get in it if he doesn't wear the veil!

Photograph by the author

WOULD YOU LIKE TO BE A BEEKEEPER?

Never too young or too old. I began beekeeping at the age of 13 and I'm writing this at 89.

Here is my first book on the subject, published by the NSW Department of Agriculture which now produces many more books on various aspects of beekeeping. Though published in 1941, some items in it are still relevant today.

The price 2/6 is 25 cents today!

And here is the record book of my first hive, 1943, still with the brown paper cover.

But you will no doubt use an electronic record.

If you are thinking of being a beekeeper, remember that there are three rules:

- If you play with bees you will get stung.
- If you can't cope with that, don't do it.
- If you are truly allergic to bee stings, don't even think about it.

Let it be emphasised, however, that bees are not necessarily aggressive, not always in an attacking mode. Bees are like people; some are nasty and aggressive, some are very docile, most are neither nasty nor especially friendly. It all depends on the personality of the queen; breeders produce queens from queen mothers chosen from quiet docile colonies. (See previous description of requeening and the change in the temperament of bees given a new queen of quiet stock.)

One beekeeper used to encourage beginners to get used to the feel of bees on the naked skin of hand and arm. Bees sitting or crawling there are not about to sting; if they are going to attack they do it like a dive bomber without warning.

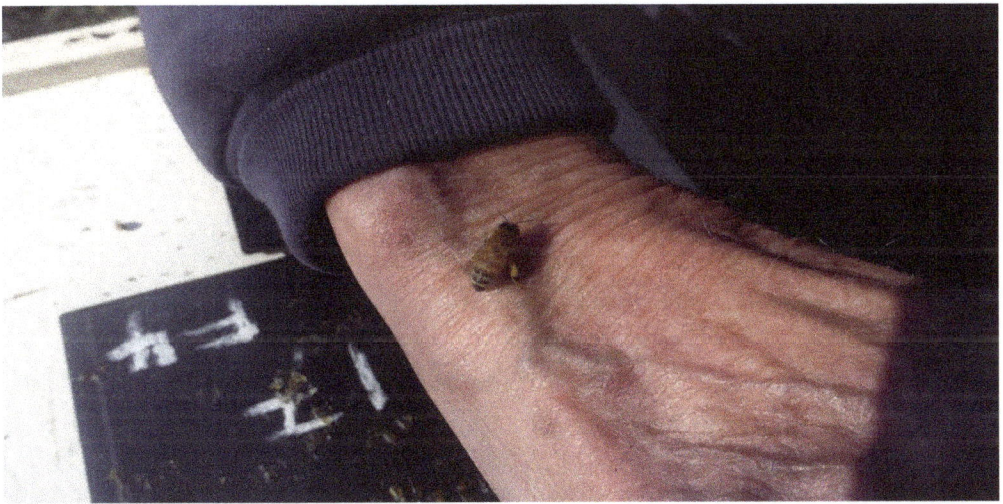

This bee is not interested in stinging, she has just returned from a foraging trip with a load of pollen in her leg baskets and all she wants to do is unload

To become a beekeeper you must :

- LEARN
- GET BEES
- GO ON LEARNING

Learning

Before getting bees you must initially learn enough to know you are committed. This means reading a book about bees, becoming aware of legal regulations pertaining to beekeeping and of the basic necessities involved in the care of bees. Then get actively involved with a beekeeper working with bees. There are two ways to do this:

- The buddy or apprentice system. One to one relationship with a friendly beekeeper, amateur or professional. (Beekeepers are a friendly lot and love to teach.)
- Join a beekeeping association, attend meetings, join in their apiary activities. The Amateur Beekeepers' Association of NSW has over 20 affiliated clubs from Bega to the far north coast. Welcoming and teaching beginners is a major objective and activity of the association.

Getting Bees

There are three ways to get bees:

- If you are lucky enough and financial enough to find a beekeeper who is down-sizing you may buy an established hive for several hundred dollars. If you do, the vendor is obliged to have inspected it and guarantee that it is free of disease.
- If you are lucky enough to happen upon a swarm on a tree in spring time, you may be able to hive it, but this needs some experience or skilled help. Sometimes a lot of help, see below.

Collecting an established swarm. You need considerable experience to collect and hive swarms, and a lot of help to tackle ones like this.

Photo by Warren Burley

Fortunately many swarms are much easier to reach and collect:

Photograph from website of the Northern Rivers Branch of the Amateur Beekeeper's Association of NSW

But you still need a little experience and the right gear.

▸ The third way to get bees is to buy a nucleus hive. This is a small hive, four or five frames, with a newly established colony produced by an experienced beekeeper by splitting a strong parent colony. Best to do this through your amateur club where you can enjoy on-going advice and support. This is perhaps the best way for a beginner to start because the new beekeeper virtually grows up with the new colony.

And when you get bees you must keep records, every time you open the hive. Record what you see and what you do. In the old days we did this with pen and paper:

But you will no doubt use an electronic record.

Go on learning… for the rest of your life

Learning about the biology of bees, about the practicalities of beekeeping are only one thing. Gaining confidence is as important. As indicated above, I began beekeeping as an early teenager but I did not get confidence until in mature years when I joined the amateur association sharing practical activities with more experienced colleagues. You have gained confidence when you can put your ungloved hand into the bottom of a hive to retrieve something you dropped there. But don't do anything like this as a beginner. Beginners should use maximum protection. Getting a walloping with stings destroys confidence.

Some Laws and rules governing beekeeping in New South Wales, Australia

Keepers of honey bees (but not Australian native bees) must register with the Department of Primary Industries and are allocated a number which must be painted or etched on to the brood box (the lowest deck of the hive). Regulations demand that the hive must be thoroughly inspected from top to bottom at least twice a year to check for disease. Good practice, however, would have you look inside the hive every month in spring, summer and autumn, leaving them alone in the cold months of winter. You can keep bees in your backyard in the suburbs if you follow a well documented code of practice and neighbours are on side. (A jar of honey over the back fence is always welcome!)

Bees make good pets. You don't have to put them in a kennel when you go on holidays and they give such sweet reward for your trouble.

Further Reading

A good start is the NSW Department of Primary Industries publication "Bee Agskills" and there are many available books on bees and beekeeping.

TRIVIA:
Some interesting statistical figures, just for fun

Figures given here are approximate only, differing from source to source. Such "rubbery" figures must not be interpreted as scientific facts, but they do give concepts of bees' activities. Don't be concerned if you detect contradicting figures!

6,600 foraging bees can collect 1kg of honey in a day

1kg of honey requires 80,000 loads of nectar

    Thus: 80 trips per gram of honey

        400 trips for a teaspoon

        40,000 for a 500g jar

        80,000 for a 1 kg jar

Trips per day: 6-13

A bee would have to travel the equivalent of four times around the world to produce a kilo of honey.

Weight per load: Pollen 6-21mg, Nectar 12-70 mg

Flowers per load: 66-178

Time per load: 6-150 minutes

Today's fact
Bees need four million flowers to make one kilogram of honey.

Worker bees weigh 81-151g, drones 196-225g, queens 178-292g

Queen's capacity to lay eggs: 1000-3000 per day, Average 1600

Wings beat at 230 beats/sec, 13,000 rpm, Bees fly 24 km/hr

About Beeswax:

It is formed from glands in the 4th-7th abdominal segments of worker bees 12-18

days old

Specific Gravity 0.96-0.97 (therefore floats on water)

Melting point 64°C, Flash point 204°C

It is fat soluble, dissolves in turps, ether, chloroform

The bees do a lot of work to make it; it takes eight times more energy expenditure

by the bees to make wax than it takes to make honey.

Honey:

Don't heat above 43°C, heat destroys both flavour and health value

At 14°C it candies quickly.

Beekeepers :

In 2010 there were 9,600 registered beekeepers in Australia

Some dates of interest:

Apis mellifera brought to Australia in 1822, to America about 1638

> "Driving along the Pacific Highway near Macksville recently, I saw a hand-written sign advertising free-range honey," Jeremy Smith, of Armidale, reports. "I wonder if readers might be able to suggest other kinds of honey worth looking out for." Suggestions must exclude honey made by battery bees.

> "Driving along the Pacific Highway near Macksville recently, I saw a hand-written sign advertising free-range honey," Jeremy Smith, of Armidale, reports. "I wonder if readers might be able to suggest other kinds of honey worth looking out for." Suggestions must exclude honey made by battery bees.

An interesting letter about in a local newspaper.

Can you think of any bees *not* free-range? Those locked up in sailing ships en route from Europe to Australia in 1822 were not free-range, but they were not making much honey in the middle of the ocean!

Cartoons by Peter Lewis reproduced with permission

ACKNOWLEDGEMENTS

Elizabeth Frost, Darren Bayley, Michelle Smith, Rod Bourke, Doug Somerville and staff of the Department of Primary Industries, Tocal Agricultural College for supply of photographs.

Des Cannon, editor of *The Australasian Beekeeper* for providing photographs, help and encouragement.

Bill Winner for supplying photographs and for providing permission of Capilano to use material for children's entertainment.

Gina Cranson, for permission to use her many artistic paintings of native bees.

Anne Creevey, member Hunter Valley Branch of The Amateur Beekeeping Association, for drawings of bee anatomy.

Anthony Pyne, professional beekeeper, for supplying photographs.

Members of the Hunter Valley Branch of The Amateur Beekeepers' Association of NSW who also provided photographs.

Peter Lewis, Newcastle artist and cartoonist, for permission to reproduce the small cartoons bearing his name. They were published in *The Newcastle Herald* over many years.

ABOVE ALL

The school kids whose lively interest displayed at the beekeepers' stalls at Tocal and Local Government Shows and in school class room talks provided the compelling inspiration to write this book.